D1387076

Design of VMOS Circuits, With Experiments

by
Robert T. Stone & Howard M. Berlin

Howard W. Sams & Co., Inc.
4300 WEST 62ND ST. INDIANAPOLIS, INDIANA 46268 USA

International Standard Book Number: 0-672-21686-8
Library of Congress Catalog Card Number: 79-57617

Printed in the United States of America.

Preface

Since the first commercial VMOS power FET was introduced in 1976 by Siliconix, Inc., this new, exciting, and, perhaps, revolutionary device has been the center of continuing controversy. The VMOSFET, or vertical metal oxide semiconductor field-effect transistor, is in many ways a giant step toward the ideal active circuit element. In the evolution of solid-state technology, VMOS devices follow the bipolar junction transistor, the field-effect transistor, and the CMOS, with an impressive array of characteristics, such as high input impedances, nanosecond switching times, low feedback capacitance, no thermal runaway or second breakdown, and high transconductance.

In spite of these characteristics, many manufacturers have developed a conservative "wait and see" attitude, while others have committed themselves to exploiting its potential. Following Siliconix, other manufacturers, including Semtech, International Rectifier, ITT, and Supertex, have produced available VMOS power devices.

In 11 chapters, this book discusses the simplicity of VMOS circuit design. It is meant to whet your appetite. You are strongly encouraged to read Chapters 1, 2, and 3 before continuing the following chapters on audio, radio-frequency, power supply, and microcomputer applications. The final chapter presents a series of laboratory-type experiments that demonstrate a number of the concepts discussed throughout this book. In addition, the appendixes include a glossary and a sampling of VMOS data sheets from several manufacturers.

We are indebted to William Roehr of Fairchild Semiconductor for his participation in the writing of the first chapter; to David Larsen without whose encouragement this project would not have been started; to Siliconix, Semtech, International Rectifier, Hitachi, and ITT for providing technical data and allowing us to reproduce the numerous data sheets found in Appendix B. Finally, we would like to thank our wives, Karen (RTS) and Judy (HMB), for their love, encouragement, and limitless patience shown during the preparation of this book.

ROBERT T. STONE
HOWARD M. BERLIN

Contents

CHAPTER 1

VMOS: A GIANT STEP TOWARD THE IDEAL 9

Introduction — Objectives — The Ideal Linear Element — The Ideal Switch — Linear Applications — Switching Applications

CHAPTER 2

THEORY OF SEMICONDUCTOR OPERATION, BIPOLAR AND FIELD-EFFECT TRANSISTORS 16

Introduction — Objectives — Conductors, Insulators, and Semiconductors — Electric Currents and Charge Carriers — Current "Hogging" — Bipolar Junction Transistor Operation — The Vertical Field-Effect Transistor — Switching Speed — Input Impedance Considerations — VMOS Disadvantages — Why Use VMOSFETs? — The HEX-FET, a Power FET — Variation in VMOS Devices

CHAPTER 3

VMOS DEVICES: APPLICATIONS AND LIMITATIONS 31

Introduction — Objectives — Handling Precautions — Power Considerations — VMOS Parameters — Protected Versus Unprotected Gates

CHAPTER 4

Basic Circuit Configurations 38

Introduction — Objectives — Common-Source Circuit — Common-Gate Configuration — Source-Follower Configuration — Switching Considerations — Inductive Switching Loads — Parallel Operation — Series Operation

CHAPTER 5

Audio Circuit Applications 51

Introduction — Objectives — Distortion — A Simple Audio Power Amplifier — A Simple 15-Watt High-Fidelity Amplifier — A 40-Watt High-Fidelity Amplifier — A 100-Watt Audio Amplifier — Class D Switching Audio Power Amplifiers — Another High-Efficiency Design: Class G

CHAPTER 6

Radio Frequency Applications 65

Introduction — Objectives — A Single VMOS 40- to 220-MHz Broadband Amplifier — A 1.8- to 54-MHz 16-Watt Broadband Amplifier

CHAPTER 7

Power Supply Applications 72

Objectives — Switching Regulators — VMOS Inverter Circuits — Linear Regulators — Overload Protection

CHAPTER 8

Microcomputer Applications 79

Introduction — Objectives — Logic-Level Interfacing — Line Driver Circuit — Peripheral Interfacing — An Example

CHAPTER 9

UNUSUAL DEVICES AND CONFIGURATIONS 88

Introduction — Objectives — A VMOS Constant Current Source — High Current Variable Resistor — Efficient Power Controls — A Fast Pulser Circuit — Analog Switches — Automotive Ignition Systems

CHAPTER 10

LARGE-SCALE INTEGRATION DEVICES 98

Introduction — Objectives — VMOS Random Access Memories — Read-Only Memory — Summary

CHAPTER 11

PERFORMING THE EXPERIMENTS 103

Objectives — Rules for Setting Up Experiments — Format for the Experiments — How Many Experiments Do I Perform? — Equipment — Components — An Introduction to the Experiments — Experiment No. 1 — Experiment No. 2 — Experiment No. 3 — Experiment No. 4 — Experiment No. 5 — Experiment No. 6 — Experiment No. 7 — Experiment No. 8 — Experiment No. 9

APPENDIX A

GLOSSARY 133

APPENDIX B

DATA SHEETS 137

INDEX 172

CHAPTER 1

VMOS: A Giant Step Toward the Ideal

INTRODUCTION

With the recent introduction of Vertical Metal Oxide Semiconductor (or *VMOS*) technology, the field-effect transistor now offers an active device to the circuit designer that possesses a combination of characteristics closer to the ideal than any other device available. Although first described by J. E. Lilienfeld in 1925, the exciting potential of the unipolar, or field-effect transistor (FET) was not realized until after the development of planar technology in the early 1960s made production practical. Since then, the MOS type of field-effect transistor has revolutionized the integrated circuit field; but discrete devices have found wide use only as rf amplifiers and mixers. Conventional FETs have a very high on-state drain-to-source voltage, rendering them unsuitable for switching significant amounts of current. In addition, they also have a "square law" transfer characteristic that rules them out for any linear, audio, or video application. The VMOS transistor overcomes these serious limitations found in FETs, and challenges the bipolar junction transistor with the following characteristics:

- High input impedance—direct interface to CMOS and TTL
- Nanosecond switching times—low inherent time delay
- Low feedback capacitance
- No thermal runaway or second breakdown
- High transconductance
- Linear transfer characteristic
- Low on-state voltage—no offset voltage

This chapter covers a basic review of the ideal active circuit element, and compares it with bipolar, FET, and VMOS devices.

OBJECTIVES

In this chapter you will learn:

- The characteristics of the "ideal" linear amplifying device.
- The characteristics of the "ideal" switching device.
- The deficiencies of previous active devices in meeting these ideals.
- The improvements in transfer characteristics offered by VMOS-FETs.

THE IDEAL LINEAR ELEMENT

An ideal linear element has a constant "gain"—that is, the relationship between a change of input signal and a change of output signal is constant regardless of the signal magnitude, frequency, or environment changes, such as temperature. To minimize the number of elements required to produce a given amplification, the gain should be high, but not so high that noise or spurious signals are troublesome.

Transfer curves showing the relationship between input and output for two types of ideal active devices are given in Fig. 1-1. If the input variable is current, then the device should have a *low* input impedance to minimize input power requirements. On the other hand, if the input variable is voltage, then the device should have a *high* input impedance. If the transfer curve is independent of out-

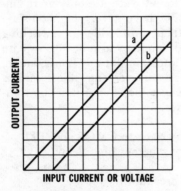

Fig. 1-1. Transfer curves for two types of ideal active devices.

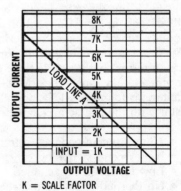

Fig. 1-2. Output characteristics relating output current and voltage for an ideal device.

put voltage, devices of either curve A or B are capable of linear amplification if the quiescent, or zero-signal point is placed midway in the transfer curve, i.e., Class A operation.

The output characteristic relating output current and voltage for an ideal device is shown in Fig. 1-2. Note that the output voltage does not affect the output current except at zero volts. Accordingly, a load line, such as line "A" can be used to achieve voltage output swings equal to the power supply voltage, which is desirable for high power efficiency.

In order for the behavior to be independent of frequency, it is required that the active device be free of any reactive (inductive or capacitive) effects or other sources of time or phase delay.

THE IDEAL SWITCH

The ideal switch has no power loss when used to interrupt current through a load, and has no limitations upon repetition rate. This simple definition imposes severe requirements: in the closed condition, or on-state, no voltage drop may exist across the device terminals. In the open condition, or off-state, no current may flow in the load. The transition between states must be performed without time delays. The output characteristic of Fig. 1-2 fulfills the first two requirements, but the third is only achieved by a device that has no reactive effects or other sources of time delays. Also of concern is the energy required to actuate the switch. Generally, the smaller the energy requirement, the better, up to the point where system noise or spurious signals may cause erroneous actuation. Those devices having transfer curves, as shown previously in Fig. 1-1, can have a zero actuation energy even though a finite control signal is required. For the current controlled device, the input impedance must be zero while an infinite input impedance is required for the voltage controlled device. Transfer curve "B" is preferable because the device is cut off when no input signal is present.

The characteristics for the ideal amplifier and switch have much in common, such as zero on-state voltage drop, high gain, and freedom from reactive effects. However, transfer curve "A" of Fig. 1-1 is preferable for a linear amplifier, while curve "B" is preferable for a switch.

LINEAR APPLICATIONS

Fig. 1-3 shows typical static transfer curves for a bipolar transistor, a standard MOSFET, and a VMOSFET compared to the ideal device. It can be readily seen that the bipolar transistor curve is the most nonlinear, and its current gain is adversely affected at low

11

signal levels by surface and volume recombination, and, in addition, at high current by loss of emitter efficiency, base-region current crowding, and base region conductivity modulation. Thus, the current gain of a bipolar power transistor can never be constant over its operating current and temperature range. The fact that bipolar devices are used in high-fidelity amplifiers is a tribute to the ingenuity of circuit designers in applying enormous amounts of local and overall negative feedback.

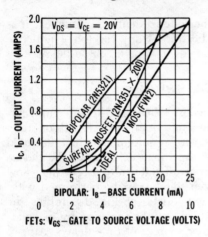

Fig. 1-3. Comparison of typical static transfer curves for bipolar junction transistors, MOSFETs, and VMOSFETs.

The transfer characteristic of a standard MOSFET is also non-linear. Its curve in Fig. 1-3 was obtained by scaling up the drain current of a 2N4351 by a factor of 200. A characteristic of "long" channel FETs is that the square root of the output current is proportional to the gate voltage. This square law characteristic generates a serious amount of second harmonic and intermodulation distortion. This problem coupled with low gains explains why conventional FETs are rarely found in audio equipment. They can, nevertheless, be used in push-pull amplifiers (Class B or AB) with acceptable results, but this connection is only attractive as a rule for power output stages.

Fig. 1-3 also shows the transfer curve of a typical VMOS transistor. Except for some curvature in the low current region, it closely approximates the ideal transfer curve "B" of Fig. 1-1. An inherent property of a "short" channel FET is that the gain is constant over most of its operating range. Low distortion is then an obvious benefit.

Since the static transfer curves are measured at a fixed output voltage, more insight into linear applications can be obtained by inspecting the output characteristics where the effect of output im-

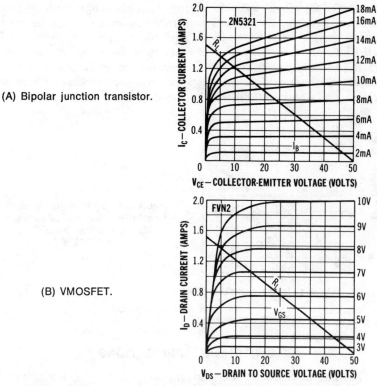

(A) Bipolar junction transistor.

(B) VMOSFET.

Fig. 1-4. Output characteristics for transistors of equal die area.

pedance is shown. Fig. 1-4 shows the output characteristics for a bipolar and a VMOS transistor of equal die area. The nonlinearities previously mentioned are evident from the nonuniform spacing of the curves. In addition, a region of gross nonlinearity exists in the low-voltage area, which is called the saturation region in bipolar transistors and the ohmic region in FETs. Obviously, load line excursions into this region will produce severe distortion. It is not obvious from Fig. 1-4 which device will produce the maximum output voltage swing without severe distortion.

The VMOS device has a higher saturation resistance, but the bipolar device is plagued with a quasi-static region at the higher current levels. Accordingly, the dynamic transfer curve obtained from those of Fig. 1-4 is shown in Fig. 1-5.

The ideal requirement that the gain is unaffected by frequency is a goal that no active device fully meets. Bipolar devices have a frequency response that is affected by a number of factors: base transit time, collector depletion layer transit time, and RC time

13

constants associated with the emitter-base and collector-base junction capacitance. Field-effect transistor capacitances can be made much smaller than those of an equivalent bipolar transistor, and all FETs are essentially free of other time delays as are discussed in the following section. The reasons for the reduction of capacitance will be readily apparent when the construction of conventional FETs and VMOSFETs is considered in a later section.

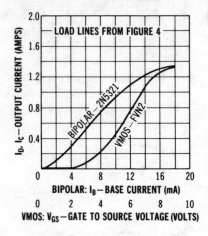

Fig. 1-5. Comparison of dynamic transfer characteristics between bipolar junction transistors and VMOSFETs.

SWITCHING APPLICATIONS

The output characteristics shown previously in Fig. 1-4 also indicate how well solid-state devices approximate the ideal switch as indicated by the graph of Fig. 1-2. The on-state voltage is higher for the VMOS transistor than for the bipolar at higher currents. Although not evident from the graph, the VMOS curves intersect at the origin and do not exhibit the offset voltage characteristics of bipolar devices, which gives VMOS an advantage at lower currents.

All types of silicon solid-state devices have satisfactorily low leakage currents when compared to the load currents being handled. However, the leakage current in a bipolar device flows out of the base, thereby demanding that a low impedance source be used, or some means be provided to reverse bias the base in order to maintain a cutoff condition. The VMOS transistor, an enhancement mode device, is cut off even with a slight forward voltage present on the gate, and has an exceptionally low gate leakage current, making it practical for direct coupling to logic outputs and for use in high impedance circuits.

The switching speed is directly related to the time delays and capacitances of the active element, as previously discussed. In addition to a frequency response limitation, the bipolar device is hampered by storage time, which is difficult to get under a microsecond in power transistors. In addition, it not only limits the switching speed, it also complicates the design of some circuits, such as power converters, by having to design a "dead time" into the drive circuits to avoid the excessive dissipation that occurs when a pair of devices in a push-pull circuit are on at the same time. All FETs are completely free of this phenomenon and when driven from low impedance sources, can switch an ampere in less than 4 ns. The switching times are primarily determined by the rate at which the device capacitances can be charged. The limiting turn-on time is on the order of 1 ns; it is simply the transit time of carriers in the channel.

In summary, VMOS power transistors are the closest electronic device to the ideal amplifier or switching device. They overcome the severe nonlinearities, frequency, and switching speed limitations of the bipolar device, and the poor linearity and high on-state voltage of conventional MOS devices. The only advantage of bipolar transistors is their superiority in the on-state voltage when used as a switch. In linear applications, the quasi-saturation region of bipolar devices limits their load line swings to about the same level as FETs.

CHAPTER 2

Theory of Semiconductor Operation, Bipolar and Field-Effect Transistors

INTRODUCTION

As one of the three forms of solid matter, semiconductors are in the middle between conductors and insulators. Because of the manner in which they conduct electricity, semiconductors fostered the development of the solid-state diode, the bipolar junction transistor, and the field-effect transistor. By changing the geometrical construction of a field-effect transistor to that of a short, v-shaped channel, the resulting vertical metal oxide semiconductor field-effect transistor, or VMOSFET, offers improved characteristics, such as faster switching speeds, high input impedance, no second breakdown, or current hogging. This chapter discusses why this is possible.

OBJECTIVES

In this chapter you will learn:

- The electrical characteristics of conductors, insulators and semiconductors.
- The characteristics of the junction diode and junction transistor.
- The operation of the vertical metal oxide semiconductor field-effect transistor.
- The advantages of VMOSFETs.
- The disadvantages of VMOSFETs.

CONDUCTORS, INSULATORS, AND SEMICONDUCTORS

All solid forms of matter may be loosely divided into three categories: conductors, insulators, or semiconductors. Certain materials are solid because outer electrons of each atom on the material are shared with neighboring atoms, thus locking each atom of the structure in place. If the outer electrons of the atom are given enough energy, such as by heating, these electrons break free from their associated atoms and are able to move easily throughout the material.

Conductors

Conductors are those materials that have a large number of outer electrons that are free to move through the material. Therefore, when a small electrical potential is applied to a conductor, such as copper, many electrons move, which, in turn, produces an electric current.

Insulators

Insulators are materials in which essentially all of the outer electrons are tightly held by the atoms in the material. When an electric field is applied to an insulator, no significant electron movement occurs, and the current flow is minimal or nonexistent.

Semiconductors

Semiconductors are materials that have a few electrons that are free from their atoms, while the rest are tightly held. As a semiconductor is heated, more electrons are freed, so that more are available to carry an electric current. When a potential is applied to a pure semiconductor, a small amount of current will flow. However, if the semiconductor is then heated, more current will flow since more electrons are now free to move within the material. The difference in these three types of materials is more a degree of free electrons than of any other feature.

Semiconductors conduct electricity in two basic ways. The first, as was described, is the movement of free electrons that have broken their attachment to their parent atoms. The atom that loses an electron now has a "hole" in the electron structure of the solid material. Neighboring electrons that do have enough energy to break loose can then easily move into the neighboring hole. If a hole occurs at the point of the material where an electron source is connected, such as the negative lead of a battery, the hole will be filled. This hole has then, in effect, allowed an electric current to flow through the material.

ELECTRIC CURRENTS AND CHARGE CARRIERS

Current can be defined as the movement of electrical charge from one place to another. The charged particle that we most often think of moving is the electron, which is a subatomic particle of matter that orbits the nucleus of the atom. Each electron has a *negative* unit of charge. Inside the nucleus of the atom are protons, or *positively* charged units. Every unaffected atom has the same number of protons and electrons, which results in a neutral charge.

When an electron is released from the pull of the atom, both the electron and the atom become charged particles, and will move in an electric field. Since the electron has a negative charge, it will move toward the positive end of the electric field. The atom, on the other hand, is positively charged and will move toward the negative end of the electric field.

In solid matter, the atoms are rigidly locked in place so that any charged atoms are unable to move about. Consequently, the major source of electric current is attributed to the movement of electrons. It has been pointed out, however, that the movement of shared electrons in a semiconductor produces a movement of holes. The effect of this action is similar to the movement of positively charged atoms.

The movement of charge carriers, or current, is produced in three basic ways. The first way is by the presence of an electrical field. Just as floating pieces of wood will be moved along by the water flow in a stream moving from a higher level to a lower level, so will electrons move in an electric field from a region of high intensity (negative source) to a region of low intensity (positive source). On the other hand, the movement of hole charge carriers is in the opposite direction, since holes are positively charged and move from positive to negative in an electric field. This type of current is known as *drift current*.

If holes and electrons are moving in the same electric field in the same region, the current produced is in the same direction. Although the holes and electrons move in opposite directions, they also have opposite charges, so that the charge transferred (i.e., current) is the sum of the hole current and the electron current.

The second type of current is the current produced by a moving magnetic field. Although we will not discuss this type in detail, the total current is still the sum of the hole and electron currents.

The third type of current produced is known as the *diffusion current*. Just as water within a container seeks its own level, the concentration of charge carriers in a semiconductor seeks to reach the same level throughout the material. If we can somehow inject electrons at one end of a bar of silicon, the electrons will move throughout the bar until the electron concentration is equal everywhere. This equilibrium is accomplished by diffusion, hence the term diffusion current.

Diffusion of charge carriers is a nondirectional, or random process. Consequently, this results in a slower transfer of charge when compared with that produced by drift currents. While the amount of drift current is proportional to the electric field, the amount of diffusion current is proportional to the difference in charge carrier concentration between the source of charge and its destination.

If we want to increase the magnitude of the diffusion current, we must increase the concentration of the charge carriers at the source, decrease the concentration at the destination, or both. In effect, we are increasing the total charge stored in the semiconductor material, which is referred to as *diffusion capacitance*. It is the difference between diffusion and drift currents that produces the remarkable differences between bipolar junction transistors (BJT) and vertical metal oxide semiconductor field-effect transistors (VMOSFET).

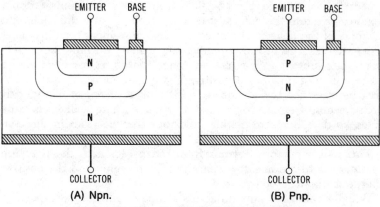

Fig. 2-1. Regions of a bipolar junction transistor.

As shown in Fig. 2-1, the current controlling region of an ordinary BJT consists of two *np* junctions. These junctions, or connections, are arranged as either' two *n* regions separated by a *p* region (an npn-type BJT), or two *p* regions on either side of an *n* region (a pnp-type BJT).

The three terminals labeled *E, B,* and *C* are the emitter, or source of current carriers; base, or current control connection; and collector, or the area where the current carriers are removed from the transistor.

It does not require an in-depth knowledge of the BJT to understand the advantages of VMOS devices over the BJT. However,

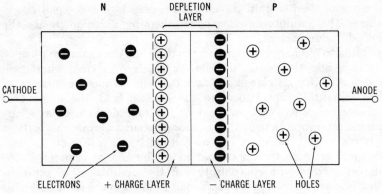

Fig. 2-2. Np diode junction.

a general knowledge of diode junction phenomena will greatly aid the reader in appreciating the origins of these advantages.

When an *np* diode junction is formed in the manufacturing process, as shown in Fig. 2-2, electrons from the neutral donor region move into the neutral *p* region (acceptor) by diffusion, and fill some of the holes in the *p* region. This movement then produces a negative charge in the *p* region, since there are now more electrons than protons in this region. Since there are less electrons in the *n* region, it now has a positive charge. In addition, the contact area between the two regions is said to be *depleted* of charge carriers, because free electrons have now moved into what used to be holes, so that there are neither holes nor free electrons in the contact area.

This movement of electrons from the *n* region to the *p* region continues until the negative charge in the *p* region and the positive charge in the *n* region restrict further movement.

When a voltage is applied across a diode with the more positive potential connected to the *p* region, or *anode,* while the negative potential is connected to the *n* region, or *cathode,* as illustrated in Fig. 2-3, the diode is said to be *forward biased.*

The applied voltage then forces electrons into the depletion region at the far end of the *n* region, while electrons are pulled out of the depletion region of the *p* region. This movement reduces the electric charge in the depletion region, while allowing electrons and holes to easily move across the depletion zone.

The applied voltage is opposite in polarity to the space-charge voltage, and produces a movement of charge carriers, or currents, through the diode.

The amount of current is determined by several factors, such as carrier doping levels, semiconductor material, size, etc., which will

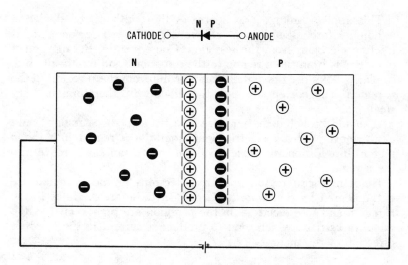

Fig. 2-3. Forward biased diode and its symbol.

not be discussed here. However, this current is also determined by the applied voltage. If the applied voltage is less than that produced by the charge in the depletion region, the diode will conduct very little, if at all. For this situation, the diode is said to be *cut off*. On the other hand, if the applied voltage is greater than this junction voltage, the resultant current through the diode increases very rapidly.

As shown in Fig. 2-4, when a positive voltage is applied to the *n* region and a negative voltage is applied to the *p* region, the diode is *reverse biased*.

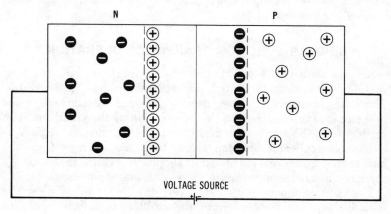

Fig. 2-4. Reverse biased diode.

This applied voltage now increases the width of the depletion region by forcing additional electrons into the p region and attracting electrons from the n region. Consequently, the movement of electrons from the n region to the p region is more difficult, and, similarly, holes are prevented from moving from the p region to the n region. The current through the diode will, therefore, be very small.

It should be noted that any free electrons in the p region could easily move with an electric field from the p region into the n region. In addition, any holes in the n region can easily move into the p region.

Holes in the n region and free electrons in the p region are outnumbered by the *majority* carriers in those areas. Holes in the n region and free electrons in the p region are then known as the *minority carriers*, which tend to produce a current in the opposite direction of the majority carriers.

CURRENT "HOGGING"

If the temperature of an np junction is somehow increased, the junction voltage decreases. If a constant voltage is applied across a diode junction, and the junction is also heated, the current through the diode increases. This current increase, in turn, produces additional heating, increasing the diode current further. This vicious cycle is known as *thermal runaway.*

When two or more np junctions are operated in parallel, one device will tend to heat slightly more than the other device and will, therefore, "hog," or draw more current. Current hogging can occur between different areas of the same diode junction, and can cause device failure as a consequence of excessive current flow and overheating.

BIPOLAR JUNCTION TRANSISTOR OPERATION

As was shown in Fig. 2-1, the bipolar junction transistor uses two diode junctions to control the current through the transistor. The base-emitter junction is normally forward biased, and serves to inject charge carriers into the base region of the transistor. The carriers from the emitter are minority carriers in the base region.

The base-collector junction serves to collect carriers from the base region, and is reverse biased. It should be recalled that minority carriers can easily move through a reverse-biased junction.

The base region must be made very thin in order to allow the minority carriers, injected from the emitter, to diffuse easily to the collector.

To switch a junction transistor from off (no collector current) to on (collector current limited only by an external resistor) requires the discharging of the base-emitter depletion region, discharging of the base-collector region, charging the base region with minority carriers, and a time delay to permit the base diffusion current to begin.

These charge changes (space-charge and diffusion capacitances) along with the time necessary for the establishment of the necessary diffusion current in the base region both limit the switching speed of the transistor.

THE VERTICAL FIELD-EFFECT TRANSISTOR

The vertical field-effect transistor, or VFET, is somewhat similar in its construction (Fig. 2-5) to the bipolar junction transistor. The terminals of a VFET are the source (S), which provides the source of current carriers through the transistor; the gate (G), which controls the formation of a current-carrying channel through the transistor; and the drain (D), where current carriers are collected.

The source of the VFET is a heavily doped n region (N+). Separating the source and drain is a very lightly doped p region (P−). When the gate voltage is zero, the charge carriers in the base region are attracted away by the junction formations at the source and drain. Thus, no current can flow through the transistor while the gate-source voltage is zero.

The drain consists of a lightly doped n region (N−) adjacent to the lightly doped p region and a heavily doped n region near the metallic drain connection.

In normal operation, as shown in Fig. 2-6, a positive voltage is applied to the gate. No current flows into the gate since the metal

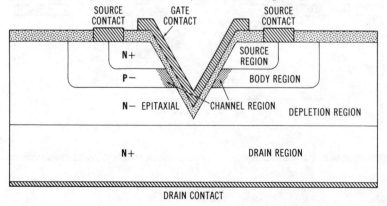

Fig. 2-5. Cross section of an n-channel VMOSFET.

gate is insulated from the body of the transistor by a layer of non-conducting material, usually silicon dioxide. Instead, electrons from the source are attracted to those areas just beneath the gate metal, just as the plates of a capacitor build up excess electrons on the negative potential plate.

If the charge on the gate is great enough, a large number of electrons will be drawn into the region next to the gate, forming an n channel region between the source and the drain. If a voltage is now applied between the source and the drain, current will flow through the channel.

The amount of current that flows through this n channel is proportional only to the number of carriers within the channel and the speed of the carriers through the channel. If a higher voltage is applied between the source and drain in an attempt to increase the speed and, in turn, the current, the carriers will then collide with electrons that are attached to the atoms in the channel region. This action produces heat and increases the resistance of the channel, which, in turn, limits any current increase.

This current-limiting feature permits two channels to operate in parallel to double the current capacity. In this case, neither channel will hog the current, since the channel which conducts more current will heat and increase its resistance until the current divides between the two channels.

Both the current-divison and limiting actions will then allow hundreds, or thousands of similar channels to be connected in parallel without current hogging. In actuality, a 2-ampere VMOS tran-

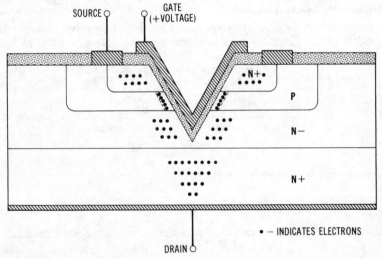

Fig. 2-6. Cross section of an n-channel VMOSFET—normal operation.

sistor is really hundreds of individual channels on a single piece of silicon that are connected in parallel. If a greater current capacity is required over that which can be obtained from a single VMOS transistor, a number of transistors may be connected directly in parallel without current-limiting resistors for current balance between VMOS transistors.

SWITCHING SPEED

In order to build more efficient power supply regulators and new classes of amplifiers, the designer must rely on high speed switching to accomplish his goal. When used for high-frequency applications, the designer was limited by the unavailability of high speed junction transistors. However, in many cases, the VMOS transistor provides us with a solution to this problem.

By utilizing the nature of the vertical groove construction, the area presented to the drain and the source is greatly reduced when compared with the standard FET (planar) construction, or with power switching transistors. This means that the capacitance between the gate and source and the source and drain can be made quite small. When this is accomplished, the device capacitances may then be quickly charged or discharged, so that typical switching speeds of 4 ns are possible for a 2-ampere VMOS transistor. By comparison, power junction transistors typically have switching speeds that are a factor of 1000 times slower.

Just what is all the excitement about switching speed? From Ohm's law, the power loss in a resistor is the product of the current through and the voltage across the resistor. The same holds true for transistors. When a transistor is off, its power is zero, since no current flows in the collector. However, when the transistor is on, very little power is lost because the transistor behaves like a small resistor so that the voltage drop across it is small.

During the time when the transistor switches from off to on, the voltage drop continually decreases from maximum to minimum with a corresponding increase in current flow from minimum to maximum. In addition, there is a loss in power during this time period so that long switching times will produce a greater energy loss. The VMOS switching transistor is much faster than its junction transistor counterpart, and can operate with much lower switching losses.

INPUT IMPEDANCE CONSIDERATIONS

If the junction transistor is operated in the common emitter mode, in which the emitter is grounded, the input impedance of the transistor is quite low, which can vary from a few ohms in power transistors to several hundred ohms for smaller types. In this configuration, the input impedance is highly dependent upon the input current, which changes with the input signal. The result of this impedance variation is a nonlinear transconductance for junction transistors. As a consequence, the nonlinear output-current input-voltage relationship greatly complicates the design of linear distortion-free amplifiers with junction transistors.

For VMOSFETs, the input impedance is, for all practical purposes, infinite. The total input current for a maximum input voltage to a typical VMOSFET is less than 0.5 μA when 15 volts dc is applied between the gate and the source. The dc input impedance is then greater than 15 V/0.5 μA, or 30 MΩ, and is not significantly changed by input current changes.

The transconductance of a VMOS transistor may be either linear or nonlinear, depending on the region of operation. For gate-to-source voltages (V_{GS}) less than 4 volts, the transconductance varies with the applied input voltage. However, for voltages greater than 4 volts, the transconductance remains relatively constant, thus simplifying the design of linear amplifiers.

The input current is limited to that necessary to charge the small gate-to-source and gate-to-drain capacitances. Consequently, VMOS transistors may be directly connected to any voltage signal source without any special interface circuitry.

The advantages of VMOS power transistors over junction power transistors are as follows:

- Greater than a 1000:1 increase in switching speed
- Higher input impedance that does not vary with input current
- Self-protection against thermal runaway and current hogging

VMOS DISADVANTAGES

By now you are probably asking yourself the question: Where's the catch? What's the gimmick? What do I give up when I use VMOS? Present users of VMOS devices for motor control design indicate that the most serious disadvantage is the long time delays in obtaining large quantities of VMOS transistors! However, the number of VMOS manufacturers is on the increase, so that this may no longer be a problem.

Seriously, VMOS devices that are currently available do have some limitations, particularly in low-voltage applications. While junction transistors may have saturation voltages as low as 0.1 volt to 0.2 volt at rated current, the saturation voltage of most VMOS transistors is on the order of 1 volt to 2 volts. As a consequence, VMOS devices exhibit a greater power loss in low-voltage circuits.

Another disadvantage is the availability of only n-channel devices in late 1978. Now several manufacturers have begun to produce p-channel devices.

Since VMOSFETs are insulated-gate devices, they are subject to damage by static electricity. Therefore, greater care is required when handling VMOSFETs. However, one manufacturer cites that the danger of handling damage is relatively small ". . . unless one is wearing crepe-soled shoes on a nylon carpet, while handling VMOSFETs near a ground."

WHY USE VMOSFETs?

When time-tested designs using a wide variety of junction transistors are readily available, why should one utilize VMOS devices in new circuit designs? The answer to this question is related to how cost effective it will be.

The inherent advantages of VMOS circuit designs are as follows:

- Simplicity
- Reduced size requirements
- Self-protection against thermal runaway

One by one, each of these advantages will be discussed in the following sections.

Circuit Simplicity

The equivalent circuit of a VMOSFET operating in the saturation region* is shown in Fig. 2-7. Shown are a linear current source and three terminal capacitances, whose typical values are as follows:

$$C_{gd} \simeq 10\,pF$$
$$C_{gs} \simeq 50\,pF$$
$$C_{ds} \simeq 40\,pF$$

For low-frequency applications, these capacitances are small enough to be neglected.

The current through the transistor is the product of the forward transconductance, g_{fs}, and the gate-to-source voltage, $V_G - V_S$. Typically, the transconductance for a 2-ampere VMOSFET is 250 mhos. This means that each 1-volt change in V_{GS} produces a 250-mA change in current. In the succeeding chapters, the model shown

---

*Refer to Appendix A to note the difference between saturation region of VMOSFETs and junction transistors.

in Fig. 2-7 will be used to design and test a number of VMOS circuits.

The simplicity of circuit design is aided by the increased linearity of VMOS as compared with standard FETs and junction transistors. Linear circuit performance using nonlinear devices requires the subtraction of the nonlinear output signals from the output of the circuit at the input (negative feedback). This is accomplished by the addition of resistors and capacitors. If the selection of feedback components, as well as their placement within the circuit is not done correctly, the circuit may tend to oscillate instead of amplify.

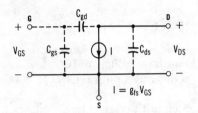

Fig. 2-7. VMOS equivalent circuit.

Circuit simplicity is also aided by the extremely high input impedance of the VMOSFET. Connections to various logic families or analog signal sources requires the application of only the input voltage drive to provide the desired output current. Since the current flow into the gate is required only to charge both the gate-to-source and gate-to-drain capacitances, the result of a direct connection to low-current logic families, such as CMOS and low-power Schottky devices, is an increase in the switching time.

Finally, when the current requirements of a given circuit are not possible with a single VMOSFET, additional devices may be operated in parallel directly *without the addition of any other components*. No current ballasting circuitry is required.

Reduced Size Requirements

The reduction in circuit package size requirements in switching applications is the result of the speed and impedance characteristics of the VMOS. Switching losses, as described earlier in this chapter, severely limit the operating frequency of junction transistor switching regulators and, thus, require significant power from the drive circuitry. As a consequence, additional heat sinking and the selection of physically large inductors and filter capacitors are required to dissipate heat and remove switching transients from the output of the regulator.

The VMOSFET permits a tenfold increase in the operating frequency, which will reduce the switching drive power and the heat sink size, not to mention the reduction in size of the filter inductor and capacitor. Therefore, if a VMOS circuit is used there will be a significant savings in physical size as well as a reduction in the complexity of the design.

Self-Protection

Thermal runaway, which is a potential problem with bipolar junction transistors, is not a factor in the design of VMOSFET circuits.

For the n-channel FET, current through the channel increases as the drain-to-source voltage increases. This increase is due to the increased velocity of electrons through the channel. As the velocity of the electrons increases, so does the number of collisions of charge carrying electrons with the electrons in the silicon crystal structure. This, in turn, reduces the velocity of the electrons. After a while, the point is reached where any further increase in the drain-to-source voltage produces no further increase in velocity or current. At this point, the channel is said to be in *saturation*.

Any increase in temperature in a VMOSFET produces an increase in the random movement of electrons in the crystal structure, and increases the likelihood of collision with charge carriers. This increase in temperature reduces the carrier velocity and decreases the current through the transistor.

Consequently, the reduction of current with increasing temperature self-regulates, or protects, the VMOSFET, allowing circuit design without temperature compensation networks, and allows for direct parallel operation for increased current capacity.

THE HEXFET, A POWER FET

As mentioned in other parts of the book, the major disadvantage of the VMOSFET in switching power supply applications is the relatively high "on" channel resistance—typically 1 ohm to 2 ohms.

International Rectifier has overcome this shortcoming in the power FET by producing a device that uses a hexagonal, multiple-source cell configuration to obtain channel resistances as low as 0.055 ohm at 28 amperes in a 100-volt device (IRF 150). International Rectifier's high-voltage offering is the IRF 350, rated at 400 volts, 11 amperes and an r_{DS} (incremental channel resistance) of 0.3 ohm.

All of the usual benefits of low input power, high speed, etc., are available with these devices, making them an ideal choice for new switching applications.

The hexagonal source cells are surrounded by channel regions, forcing current to flow laterally, then vertically from the source to the drain. Fabrication using large-scale integrated circuit techniques results in a tremendous increase in channel area, and the reduction in r_{DS} "on."

VARIATION IN VMOS DEVICES

In August 1979, Intersil announced a complete line of 15 vertical MOSFETs with r_{DS} "on" of 0.5 ohm, I_D of 4.0 amperes to 5.0 amperes, and breakdown voltages from 40 volts to 80 volts.

Intersil offerings feature a v-groove with a flat bottom, which is claimed to increase device stability.

CHAPTER 3

VMOS Devices:
Applications and Limitations

INTRODUCTION

In this chapter, you will learn the terminology used in conjunction with VMOS devices, how to select the appropriate device for a given application, some handling precautions for VMOS devices, and the characteristics of VMOS devices in different modes of operation. This chapter also will present a typical data sheet and discuss how to extract useful information from it.

OBJECTIVES

In this chapter you will learn:

• Handling precautions for using VMOSFETs.
• The definitions of VMOS parameters.
• The meaning of information given on a typical VMOSFET data sheet.
• The difference in protected gate and unprotected gate VMOS-FETs.

HANDLING PRECAUTIONS

Currently manufactured VMOSFETs are supplied either with static protected gates or unprotected gates. As discussed in Chapter 2, a gate is insulated from the body of the transistor by a thin layer of silicon dioxide. However, the gate insulator can be punctured by a high voltage between the gate lead and the body of the transistor. Once the insulator is damaged, the device is then useless.

The *protected* gate VMOSFET has a zener diode fabricated into the structure during the manufacturing process. Typically, voltages between ground and +10 volts may be applied between the gate and source (V_{GS}) without any noticeable effect on the zener diode. If the gate voltage exceeds the zener voltage, typically 15 to 18 volts, the zener conducts heavily, effectively shorting the high gate voltage source.

However, the zener is fabricated in such a manner so that the gate acts like the emitter of an npn transistor if the gate voltage becomes negative with respect to the source (positive, for *p*-channel transistors). If this occurs, the transistor conducts heavily, destroying itself. For this reason, *protected gate VMOSFETs are not suitable for those applications where the gate voltage may become reverse biased.*

If you have ever had the misfortune to damage a MOSFET by incorrect handling, it is understandable that you might be extremely cautious when using VMOSFETs. While it is possible to damage unprotected gate devices by careless handling, the chance of your doing so is remote. Studies relating the rate of static discharge into a capacitor indicate that the dielectric strength and static energy levels, found under normal handling, make the probability of damage practically nil.

When synthetic fabrics are worn, or when standing on carpets made from synthetic materials, one should take the precaution of either wearing an antistatic wrist strap, or connecting a bare wire from one's wrist watch to a ground in order to prevent any static damage to the device.

In Chapter 11, you will have the opportunity to perform a number of experiments utilizing VMOSFETs. Since these have very high input impedance and speed capabilities, you may discover how easy it is to build a uhf oscillator capable of destroying the VMOS device you are testing!

Long wire leads provide sufficient inductance and capacitance to form resonant circuits to permit high-frequency oscillation. Ordinary resistive loads, or even large inductive loads may possess enough stray capacitance to allow a large enough high-frequency current to exceed the maximum power rating. Therefore, breadboard leads must be kept as short as possible, while installing either a ferrite bead or a 100- to 1000-ohm resistor in series with the gate.

POWER CONSIDERATIONS

The majority of VMOSFETs are power devices. Because of this, power supplies capable of delivering 3 amperes and voltages up to 50 volts are normally required for experimentation.

Adequate heat-sinking of transistors is required for high-power types. Unfortunately, there are no quick and easy methods that predict the behavior of the many heat sinks that are commercially available.

VMOS PARAMETERS

The operating characteristics of VMOS and MOSFET devices have similar names and meanings. For purposes of illustration, the following specifications are typical for the model VN46AF n-channel VMOS power FET manufactured by Siliconix.

1. Absolute Maximum Ratings
   - Maximum drain-source voltage, V_{DS} (40 V)—greater voltage can cause current avalanche within the transistor. The maximum current will no longer be limited by velocity saturation, and device destruction may result from overheating.
   - Maximum drain-gate voltage, V_{DG} (40 V)—greater voltage can cause damage to the insulating layer between the metal gate and the body of the transistor.
   - Maximum continuous drain current, I_D (2.0 A)—currents greater than this maximum may overheat and destroy the transistor.
   - Maximum pulsed drain current, I_D (3.0 A)—pulsed currents greater than this value may create thermal spikes which can overheat and destroy the transistor.
   - Maximum continuous forward gate current, I_{GF} (2.0 mA)—this is the maximum sustained current into the gate protection diode without exceeding its power capacity.
   - Maximum pulsed forward gate current, I_{GF} (100 mA)—this is the maximum pulse current into the gate protection diode without the risk of overheating and destruction.
   - Maximum continuous reverse gate current, I_{GR} (100 mA)—this current limit prevents excess power dissipation in the effective npn transistor formed by the protection diode and the drain.
   - Maximum forward gate-source (zener) voltage, V_{GS} (15 V)—gate-source voltages in excess of this amount will produce avalanche currents in the gate-source protective diode, and could overheat and destroy the device.
   - Maximum reverse gate-source voltage, V_{GSR} (0.3 V)—reverse

gate-source voltages in excess of this value may produce a junction transistor effect between the protective diode and the drain. This applies only to diode-protected gate devices.
- Maximum dissipation at 25°C case temperature, P_D (12.5 W)—when the case temperature is 25°C, this power dissipation is the maximum allowable to prevent overheating the active region of the transistor.
- Linear derating factor (100 mW/°C)—the maximum power dissipation must be reduced as the case temperature increases by each 1°C. The maximum power dissipation for a case temperature of 50°C would be

$$12.5 \text{ W} - (100 \text{ mW/°C})(50° - 25°) = 10.0 \text{ W}$$

- Operating and storage temperatures, T_J and T_{stg} (−40 to +150°C)—this is the maximum temperature range to prevent excessive mechanical stress in the transistor package.
- Lead temperature, $\frac{1}{16}$ inch from the case for 10 seconds (300°C)—this value gives soldering information. Excessive temperatures, or time, may overheat the transistor and destroy it.

2. Electrical Characteristics—Static
- Drain-source breakdown voltage, BV_{DSS}—the minimum drain-source voltage that will produce a specific drain current when the gate-source voltage is zero.
- Gate threshold voltage, $V_{GS(th)}$—the intercept of the drain current (I_D) versus gate-source voltage (V_{GS}) transfer curve when the linear region is projected onto the V_{GS} axis. All linear designs require operation above this gate threshold level.
- Gate body leakage, I_{GSS}—this is the range of current that should be anticipated when the normal maximum gate voltage (e.g., 10 V) is applied. The maximum current (e.g., 100 nA) implies a minimum gate-body leakage resistance of 100 MΩ.
- Zero gate voltage drain current, I_{DSS}—the drain current when the gate-source voltage is zero.
- On-state drain current, $I_{D(on)}$—the drain current when the gate-source voltage is 10 V, and the drain-source voltage is specified (e.g., 25 V).
- Drain-source saturation voltage, $V_{DS(on)}$—this is the product of the "on" channel resistance r_{DS} and the drain current I_D.

3. Electrical Characteristics—Dynamic
- Forward transconductance, g_m—the ratio of the drain current to the gate-source voltage less the threshold voltage, so that

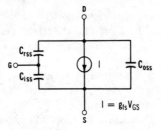

Fig. 3-1. VMOS dynamic equivalent circuit.

$$g_m = \frac{I_D}{V_{GS} - V_{GS(th)}}$$

- Input capacitance, C_{iss}—this is the sum of the gate-source and gate-drain capacitances.
- Reverse transfer capacitance, C_{rss}—this is the drain-to-gate capacitance which results in signal feedback.
- Common-source output capacitance, C_{oss}—the effective capacitance between the source and drain when the transistor is operated in the common-source mode. The dynamic model of a power VMOSFET, showing C_{iss}, C_{rss}, and C_{oss} is given in Fig. 3-1.
- Turn-on delay time, $t_{d(on)}$—the time in nanoseconds between

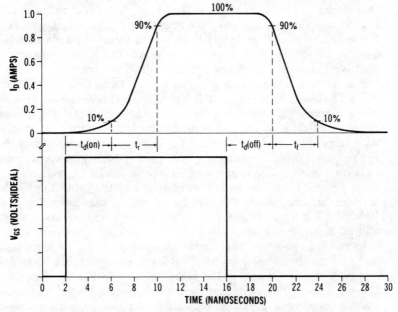

Fig. 3-2. Delay time, rise time, and fall time relationships.

the application of the gate voltage signal and when the drain current rises to 10% of its final value.

- Rise time, t_r—the time required for the output current to increase from 10% to 90% of its final value after the application of a gate voltage "on" pulse.
- Turn-off delay time, $t_{d(off)}$—the time from the removal of the gate "on" voltage until the drain current falls to 90% of the "on" value.
- Fall time, t_f—the time required for the output current I_D to decrease from 90% to 10% of the "on" value after the removal of the gate input signal.

Fig. 3-2 shows the relationships of $t_{d(on)}$, t_r, $t_{d(off)}$, and t_f versus V_{GS}.

PROTECTED VERSUS UNPROTECTED GATES

The danger of damaging the gate-body insulator by overvoltage can be greatly reduced by the fabrication of an avalanche diode into the body of the transistor during manufacture. If the cathode of this diode is connected to the gate, while the anode is connected to the source for a p-channel device, no gate current will flow when normal gate voltages are applied.

If the positive gate voltage exceeds the breakdown voltage of the diode, typically 15 volts to 18 volts, the diode will conduct heavily, thus preventing any further increase in gate voltage. Since the breakdown voltage is less than the maximum voltage that the gate insulation can withstand (40 volts or more), the transistor is then well protected from low-current voltage sources.

Fig. 3-3 shows the cross section of a protective diode in a VMOS-FET. It should be noticed that the diode actually forms an npn junction transistor with the body of the VMOSFET. During normal operation, this is of no consequence since the emitter is positive with respect to the base, so that the transistor is cut off. However, if the gate should go negative, the emitter-base junction is now forward biased and the npn transistor will conduct heavily.

To prevent excess current from flowing into the protective diode, a series limiting resistor should be placed in the gate lead of a VMOSFET if there is any possibility that the gate-source voltage will go negative (positive for p-channel devices).

In normal operation, the equivalent circuit of the protective diode and the body of the VMOSFET are shown in Fig. 3-4. The "support" npn transistor is the body of the VMOSFET. Since the N+ source and the p-channel are electrically connected, the base-emitter junction is permanently biased *off*. The support transistor is shown only to demonstrate that excessive voltages or heating

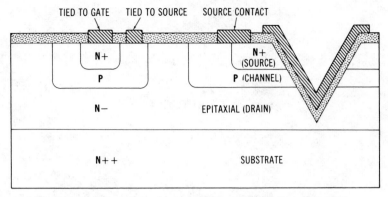

Fig. 3-3. Cross section of a monolithic VMOS gate zener diode.

can produce an avalanche breakdown in the body of the VMOS-FET.

The gate protection "diode" transistor is normally operated with the base-emitter junction reverse biased, so that transistor action occurs only when the gate-source voltage is reversed.

Unprotected gates should be handled with care to prevent static damage to the gate insulation. The best handling technique is to store the device in a conductive wrapping, such as aluminum foil, or insert the leads into a conductive foam until ready for installation. When all other circuit connections are complete, the VMOS device should be installed after making sure that the circuit, tools, your body, and the case of the device are all at the same potential.

The decision for selecting an unprotected gate versus a protected gate device is determined by several factors. Situations where the VMOS device may be handled by relatively untrained personnel require protected gate devices whenever possible. Circuits that require minimal input capacitance, or which can allow the gate-source voltage to become reverse biased, must specify *unprotected* gate devices.

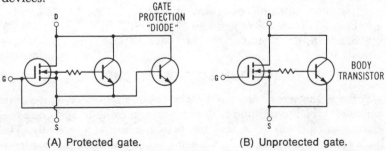

(A) Protected gate.   (B) Unprotected gate.

Fig. 3-4. Equivalent circuit of the protective diode and body of a VMOSFET.

CHAPTER 4

Basic Circuit Configurations

INTRODUCTION

This chapter discusses the three basic circuit configurations in which VMOSFETs are used, along with the performance characteristics of each: common-source, common-gate, and common-drain, or source follower. In addition, you will learn how to increase the voltage and current capacities by parallel and series connections.

OBJECTIVES

In this chapter you will learn:

- The wiring and performance characteristics of the
  a. common-source configuration.
  b. common-drain configuration.
  c. common-gate configuration.
- Parallel device characteristics of VMOSFETs.
- Series device characteristics of VMOSFETs.

COMMON-SOURCE CIRCUIT

Perhaps the most common VMOS circuit is the *common-source* amplifier, which corresponds to the common-cathode and common-emitter circuits for tube and transistor circuits, respectively. The basic circuit diagram and equivalent circuit model for the common-source configuration are shown in Fig. 4-1.

Since no input current flows for low frequencies, the current gain of the amplifier is quite high, approaching infinity. The voltage gain is the product of the forward transconductance g_m (sometimes referred to as g_{fs}) and the parallel combination of the load resistance R_L and the incremental channel resistance r_{DS}, so that

$$A_v = g_m \frac{R_L r_{DS}}{R_L + r_{DS}} \qquad (\text{Eq. 4-1})$$

where,
  A_v is the voltage gain,
  g_m is forward transconductance,
  R_L is load resistance,
  r_{DS} is incremental channel resistance.

To illustrate the design of a VMOS Class A amplifier, we will consider the following example.

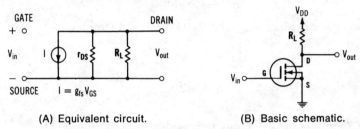

(A) Equivalent circuit.   (B) Basic schematic.

Fig. 4-1. Common-source configuration.

We want to design a Class A amplifier using a 2N6657 VMOS protected gate transistor. The circuit design is to take into account the following specifications:

Input signal: 1.67 volts peak-to-peak, 50 kΩ source impedance, 20–20,000 Hz
Voltage gain: 6, for a 25-ohm load
Supply voltage: 24 volts

The input power is then calculated as

$$P_i = \frac{V_{rms}^2}{R_S}$$

$$= \frac{(1.67 \times 0.707)^2}{50 \text{ k}\Omega}$$

$$= 0.0000278 \text{ watt or } 28 \text{ }\mu\text{W}$$

and the power output is

$$P_o = \frac{V_{rms}^2}{R_L}$$

$$= \frac{(7.07)^2}{24}$$

$$= 2.08 \text{ watts}$$

so that the power gain is 75,000.

The bias point is selected at the midpoint of the supply voltage, so that the drain-source voltage $V_{DS} = 12$ volts. Consequently, the drain current with no input signal is

$$I_D(0) = \frac{V_{DS}}{R_L}$$

$$= \frac{12}{24}$$

$$= 0.5 \text{ ampere}$$

The power dissipated by the VMOSFET is, therefore,

$$P_D = I_D V_{DS}$$

$$= (0.5)(12)$$

$$= 6 \text{ watts}$$

The data sheet for the 2N6657 gives a power derating of 200 mW/°C so that the maximum temperature allowable is

$$T_{max} = 25 + \frac{P_D - 6\,W}{0.2\,W/°C}$$

$$= 25 + \frac{25 - 6}{0.2}$$

$$= 120°C$$

Since r_{DS} is usually much greater than the load resistance of 24 ohms, the voltage gain is then

$$A_v \simeq R_L g_m$$

$$= (24)(0.250)$$

$$= 6$$

To prevent input signal loss due to bias circuit shunting effects, the input impedance of the amplifier should be at least 10 times the impedance of the signal source, or at least 500 kΩ. For the 2N6657, the I_D versus V_{DS} characteristic curve shows the gate voltage V_{GS} should be approximately 4.3 volts, if the drain current and drain-source voltage are to be 0.5 ampere and 12 volts, respectively. Using the input bias circuit shown in Fig. 4-2, the bias resistors are computed using the pair of simultaneous equations:

$$R_{input} = \frac{R_A R_B}{R_A + R_B} \geqq 500 \text{ k}\Omega \qquad (\text{Eq. 4-2})$$

and

$$V_{GS} = (24 \text{ V})\frac{R_B}{R_A + R_B} \qquad (\text{Eq. 4-3})$$
$$= 4.2 \text{ volts}$$

A suitable pair of resistors would be 3.9 MΩ and 820 kΩ for R_A and R_B, respectively.

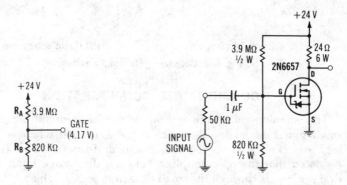

Fig. 4-2. Gate bias circuit.   Fig. 4-3. Common-source amplifier.

For a Class A amplifier, the input-coupling capacitor should be chosen so that its impedance, X_C, is *less* than the input impedance of 500 kΩ at 20 Hz (the lowest input frequency), so that

$$C = \frac{1}{(6.28)fX_C}$$
$$= \frac{1}{(6.28)(20 \text{ Hz})(500 \text{ k}\Omega)}$$
$$= 0.000000016 \text{ farad or } 0.016 \text{ } \mu\text{F}$$

Selecting a value of 1.0 μF will ensure that the loss of gain at 20 Hz will be minimal. The completed circuit is shown in Fig. 4-3. It should be noted that the actual power supplied by the signal source is then

$$P_S = \frac{(1.67 \times 0.707)^2}{50\,\text{k}\Omega + 677\,\text{k}\Omega}$$
$$= 0.0000019 \text{ watt or}$$
$$= 2\,\mu\text{W}$$

The common-source configuration is used where both voltage and current gain are desired. The input impedance is high along with the incremental output impedance.

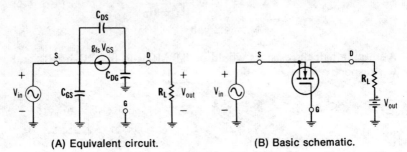

(A) Equivalent circuit. (B) Basic schematic.

Fig. 4-4. Common-gate configuration.

COMMON-GATE CONFIGURATION

The common-gate configuration, whose basic circuit diagram and small signal model are shown in Fig. 4-4, is used whenever a very high isolation between the source and drain is required. It should be noted that the only coupling between the source input and the drain output is through the small capacitance C_{DS}, which is normally less than 50 pF.

Corresponding to the grounded-grid and grounded-base amplifiers for tubes and transistors, the common-gate amplifier cannot produce current gain since it operates as a transfer resistance amplifier. The input impedance will be the input voltage divided by the input current of the source. Since the output drain current is the same, in magnitude, as the input source current,

$$I_i = -g_m V_i$$

or

$$= -g_m V_{GS} \qquad (\text{Eq. 4-4})$$

where,

    I_i is input current,
    g_m is transconductance,
    V_i is input voltage,
    V_{GS} is gate-source voltage.

so that the input resistance is

$$R_i = \frac{V_i}{I_i}$$

$$= \frac{1}{g_m} \qquad \text{(Eq. 4-5)}$$

which results in an input resistance of about 4 ohms when the device operates in the linear transfer region.

For the common-gate configuration, the voltage gain is the same as given by Equation 4-1 for the common-source amplifier.

Common-gate circuits are found in switching applications. When VMOSFETs are used to switch analog signals or power, the drain may be isolated (open circuited) by placing the gate at the same potential as the source. When the gate is then heavily forward biased with respect to the source ($V_{GS} \cong 10$ V), the drain-to-source resistance drops to approximately 4 ohms or less, and the "switch" is closed.

SOURCE-FOLLOWER CONFIGURATION

The source-follower circuit configuration corresponds to the transistor emitter follower, or the cathode follower for vacuum tubes. Consequently, the voltage gain is unity. The current gain, on the other hand, approaches infinity for the VMOSFET circuit. Fig. 4-5 shows the schematic diagram and equivalent circuit for the VMOSFET source follower.

The major application of the source follower is in Class AB push-pull power amplifiers. For the simplified circuit shown in Fig. 4-6, the previous stages provide the necessary voltage gain, while the outputs of the two source followers act as low output impedance voltage sources to the load. Only one VMOSFET stage supplies current to the load at a given time, allowing power to be delivered to the load in excess of the output rating of either device.

In addition, it should be noted that the output signal is not inverted with respect to the input (no phase shift). When this am-

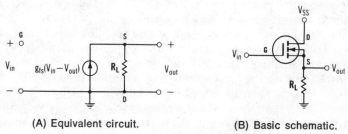

(A) Equivalent circuit.　　　　(B) Basic schematic.

Fig. 4-5. VMOS source-follower.

plifier is biased so that some current flows through both transistors (but not through the load) with no input signal, an amplifier with inherently low distortion is realized. A single-stage source follower using VMOSFETs should typically have less than 1% total harmonic distortion. In effect, this permits the construction of an audio power amplifier with less overall feedback, while possessing a reduction in transient harmonic distortion.

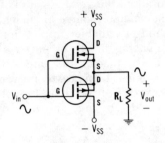

Fig. 4-6. Class AB push-pull amplifier.

SWITCHING CONSIDERATIONS

The switching speed limit for VMOSFETs is typically less than 4 ns. Turning a VMOSFET "on" requires that the charge carriers from the source region be attracted to the channel region under the metal gate electrode. For an *n*-channel device, it is "on" while allowing current to flow from the source to the drain when enough electrons are attracted to the channel to form an inversion layer. An *inversion layer* is a region where a large number of minority carriers have been deposited so that they outnumber the majority carriers. Consequently, the channel region is *inverted* from its normal carrier mechanisms, so that a *p*-type body of a VMOSFET now becomes an *n*-type current carrier in the channel.

Every electron that is supplied to form the inversion layer requires removal of an electron from the metal gate lead. This represents the gate-source capacitance C_{GS}. The rate at which this capacitance charges and, therefore, the turn-on and turn-off speeds are limited by the gate current drive source and by the breakdown voltage of the gate insulator. For fast switching speeds, the gate drive circuit must have high current sinking and current sourcing capabilities.

Since both voltage and current must be present simultaneously for power dissipation, the ideal switch would, therefore, have an infinite resistance when open, and no resistance when closed. In addition, the time to switch open or closed is instantaneous. While VMOSFETs do not achieve these ideal characteristics, the switching speed and "off" resistance are much closer to the ideal than are obtained with bipolar junction transistors.

Since VMOSFETs possess greater power losses in the "on" state, reduced switching losses, freedom from second breakdown, and reduced control power, the following conclusions can be stated:

1. The total power loss (switching plus "on" state losses) is lower than for junction transistors by operating VMOS devices at higher frequencies than are available with junction transistors. This increased frequency response reduces the necessary size of filter inductors and capacitors. This, in turn, reduces the overall size of the power controller.
2. The reduced drive power allows the use of low power analog and digital logic circuits to control the power controller, which increases the simplicity of the design.
3. The total number of components required is reduced, which, in turn, means lower manufacturing costs and increased reliability.

For proper switching, the most important consideration is to make sure that the gate-source voltage of the device is in the non-saturated (ohmic) region. However, this will depend on the load resistance and output current of the VMOSFET. Furthermore, the source impedance that is connected to the input gate should be as low as possible to ensure fast switching speeds.

As an example, a 2N6660 n-channel VMOSFET (Semtech)* is to be used to switch 1 ampere of current to a 24-ohm load within 150 ns. In addition, the output duty cycle should be 1% with an "on" time of 50 μs.

For a gate-source voltage of 10 volts, the gate-drain resistance is specified at 2 ohms. In order for the device to switch the 1-ampere current through the load *plus* the gate-drain resistance, or 26 ohms, the supply voltage must be 26 volts.

The data sheet also specifies that the device has a 10-ns delay time and a 10-ns rise time when a 25-ohm gate supply impedance is used, which is well within the specified switching time for our design. The "on" power dissipation is then

$$P = I^2 R_{GD} D \qquad \text{(Eq. 4-6)}$$
$$= (1\,A)^2 (2\,\Omega)(1\%)$$
$$= 0.02 \text{ watt}$$

where,
P is "on" power dissipation,
I is current,
R_{GD} is gate-drain resistance,
D is duty cycle.

---

* The data sheet for the 2N6660 is given in Appendix B.

so that no heat sinking should be necessary. In fact, the same results are obtained with an FVN0418Z FET housed in a TO-92 plastic case. The junction-to-air thermal resistance is specified at 200°C/W, so that the total temperature rise above the ambient temperature will be (200°C/W) (0.02W), or 4°C.

Since the switching speed and delay time are directly proportional to the gate voltage source resistance, the specified 25-ohm resistance can be increased to as high as

$$R_S\text{max} = \frac{t_s}{t_r + t_d}(R_{\text{specified}}) \qquad (\text{Eq. 4-7})$$

$$= \frac{(150 \text{ ns})}{(20 \text{ ns})}(25 \ \Omega)$$

$$= 187.5 \text{ ohms}$$

and still be able to switch the specified current in 150 ns. If the delay time can be ignored, since it will be present on both the input and output waveforms, the input source impedance can then be as high as 375 ohms, which can be realized with CMOS buffers, as shown in the completed design circuit of Fig. 4-7. On the other hand, if this circuit is to be pulsed repeatedly, the pulse generator could be an integrated circuit timer, such as the 555 timer.

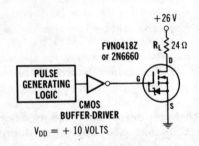

Fig. 4-7. CMOS buffer interface.

INDUCTIVE SWITCHING LOADS

The rapid switching of inductive loads generates large voltage transients called *inductive "kick."* For the majority of junction switching transistor circuits, this kick is eliminated by placing a diode in parallel with the inductor. The energy stored in the magnetic field of the inductor produces a current that flows through the diode when the transistor turns off. Consequently, this action prevents the production of the large transient voltages that could produce avalanche breakdown on the transistor.

Bypass diodes are similarly used with VMOSFETs. However, if fast devices are used, fast switching diodes are also required.

In addition, the length of unshielded leads between the VMOSFET and the load should be as short as possible, since they tend to act as indicators which could, in turn, generate a significant voltage transient at VMOS switching speeds.

PARALLEL OPERATION

One often discovers that the design that has taken many hours in development results with less than one hoped for. A corollary to Murphy's law states: "No matter what the capacity of your power supply is, the current demand will increase to the point where it just exceeds the capacity." For many VMOS designs, increased current capacity does not mean that one must repeat the design process. In fact, two or more VMOS devices may be directly connected in parallel with the original device with virtually no circuit changes.

As mentioned previously in Chapter 2, ordinary junction transistors exhibit the phenomenon known as current hogging, produced by base current distributions and base-emitter voltage drops as the temperature increases. When a second device is placed in parallel, power-robbing ballast resistors must be placed in series with the emitter leads, or else one device will hog the current, and will possibly overheat and destroy itself, the other transistor, and possibly the load.

Since current hogging is not a factor in VMOS circuits, direct parallel operation *is* permitted, keeping in mind the following considerations:

1. Some means of suppressing parasitic oscillations should be included. This can be accomplished by using ferrite beads on the gate leads, or a small resistance (100 ohms to 1000 ohms) should be placed in series with each gate.
2. Ensure that the gate drive impedance is low enough to switch all the paralleled devices at the required speed. The switching time increases directly with the number of devices in parallel (unless the driving impedance is decreased).
3. The transconductance of the paralleled devices will be the product of the single device multiplied by the number of devices that are in parallel.
4. The "on"-channel resistance will be the parallel combination of the individual devices' "on"-channel resistances.

The need for parallel operation may eventually be eliminated when wider ranges of VMOSFETs are produced. At the present time, power devices capable of switching 16 amperes at 80 volts are already available in standard TO-3 style packages.

SERIES OPERATION

The need for higher drain-source voltage capacity than is presently available has given rise to a novel technique that uses two VMOSFETs in series. This is possible because the drain current must be present for the device to enter the breakdown region.

Refer to the circuit shown in Fig. 4-8. In the "off" state, the gate voltage of Q_2 is approximately 62.3 volts, set by the voltage divider resistors R_1 and R_2. With Q_1 off, the supply voltage will be about equally divided between both transistors.

When Q_1 is turned on, the source voltage of Q_2 drops and turns Q_2 on as well. The gate of Q_2 will be held sufficiently positive by the 15-volt supply when the drain voltage of Q_2 drops, after both transistors are on. Both the 39-pF and 47-pF capacitors act as "speedup" capacitors during transistor turn on and turn off.

As the drain voltage of Q_1 rises during turn off, the source voltage of Q_2 also rises, turning Q_2 off. In effect, an equal voltage division is maintained across both transistors.

By properly selecting the gate bias voltages, any number of VMOS devices may be used in series. Large-value resistors can then be used to minimize bias power dissipation, while the individual gate voltages should be set at the desired source voltage during the off state.

As an example, the gates of transistors Q_2 and Q_3, shown in the circuit of Fig. 4-9, should be set at approximately +60 and +120 volts, respectively. Choosing R_1 as 1 MΩ for a starting point, R_2 can then be determined from

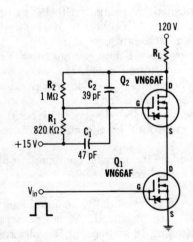

Fig. 4-8. Series operation of VMOSFETs to increase drain-source voltage capacity.

Courtesy Siliconix, Inc.

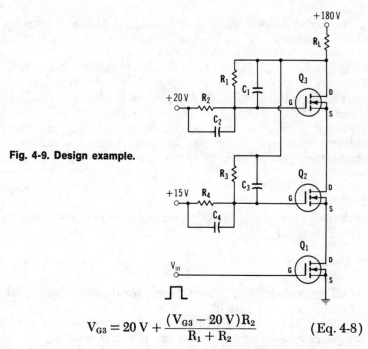

Fig. 4-9. Design example.

$$V_{G3} = 20\text{ V} + \frac{(V_{G3} - 20\text{ V})R_2}{R_1 + R_2} \qquad \text{(Eq. 4-8)}$$

and solving for R_2 gives 1.67 MΩ, for which a standard 1.5-MΩ resistor can be used.

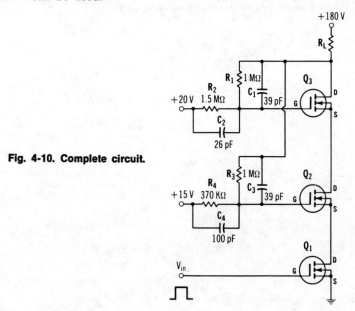

Fig. 4-10. Complete circuit.

In similar fashion, R_3 is selected to be 1 MΩ, and from the following equation,

$$V_{G2} = 15\,V + \frac{(V_{G2} - 15\,V)R_4}{R_3 + R_4} \qquad \text{(Eq. 4-9)}$$

so that R_4 is found to be 375 kΩ, for which a 390-kΩ resistor can be used.

Finally, the capacitors are chosen so that the same charge is present on each pair for good dynamic balance of the switching transistors. For one pair, we require that

$$\frac{C_1}{C_2} = \frac{R_2}{R_1} \qquad \text{(Eq. 4-10)}$$

Selecting C_1 equal to 39 pF, Equation 4-10 requires that C_2 equal 26 pF. Similarly,

$$\frac{C_3}{C_4} = \frac{R_4}{R_3} \qquad \text{(Eq. 4-11)}$$

If C_3 also equals 39 pF, then C_4 must equal 100 pF, giving the final circuit shown in Fig. 4-10.

High-voltage power FETs are already appearing from one manufacturer. When devices such as the IRF 300 and IRF 305 (International Rectifier) are used to switch up to 400 volts at 4.0 amperes and 5.0 amperes, respectively, either one can control in excess of a kilowatt of power with microwatts of drive.

CHAPTER 5

Audio Circuit Applications

INTRODUCTION

In this chapter, the principles of several designs of VMOSFET audio amplifiers will be discussed, offering inherently reduced distortion, reduced circuit complexity, and ease of self-protection circuit design. In addition, it is possible to design higher class amplifiers, such as Class D switching, or Class G amplifiers. In effect, higher class amplifiers are more efficient, which, in turn, reduces power dissipation requirements in output devices and heat sinks.

OBJECTIVES

In this chapter you will learn:

- Some of the sources of distortion in junction transistor amplifiers.
- How this distortion may be reduced or eliminated by using VMOSFET designs.
- How component parts count can be reduced using VMOSFETs in audio amplifiers.
- How VMOSFETs open new opportunities for high efficiency classes of amplifiers.

DISTORTION

VMOSFETs may very well be the answer to the audio purist's dreams. For many years, audiophiles who once were familiar with the "smoothness" of high-quality vacuum tube amplifiers have complained of a "difference" in the sound emitted from transistor amplifiers.

The performance of modern transistor power amplifiers is a credit to the designers, who have taken inherently nonlinear devices and developed low distortion amplifiers. However, music, unlike pure tones, is made up of complex nonsinusoidal components having many fast transients, particularly when percussion instruments are included.

The technique used to achieve low distortion in transistor amplifiers is to design the amplifier with more gain than desired, and then reduce the gain to the desired level using negative feedback. Negative feedback subtracts a portion of the output signal from the input signal. Consequently, most of the distortion produced in the amplifier is reduced by the same factor that the gain is reduced.

Unfortunately, reactive components, such as inductors and capacitors, are used in the feedback path of this amplifier, tending to produce slightly different responses to different input frequencies. For fast transients, the distorted output may appear before enough of the feedback signal has been returned to correct the signal and compensate for the inherent distortion of the amplifier. This distortion in amplifiers consists of the production of additional signals that are harmonic multiples of the input, which is referred to as *transient harmonic distortion*. However, this type is extremely difficult to measure, and, therefore, is rarely included in the specifications for audio amplifiers.

Since VMOSFETs exhibit a more linear transfer characteristic than junction transistors, they are ideally suited for use in audio amplifiers. In addition to their reduced distortion performance, VMOSFETs are not subject to thermal runaway and self-destruction as is the case with junction transistors. Consequently, the requirements for temperature compensation diodes in the bias circuit and power robbing emitter resistors are eliminated.

A SIMPLE AUDIO POWER AMPLIFIER

For specific audio amplifier designs, we shall first consider a low-cost alternative to present transistor circuits for use in small radios, phonographs, and televisions. Shown in Fig. 5-1 is a 2-transistor amplifier capable of supplying 4 watts to an 8-ohm load with a maximum of 2% distortion over the 100-Hz to 15-kHz range. The cost of the active devices is quite low: the VN66AF is a plastic encased power transistor which currently sells for about $2.00. While the efficiency of the circuit is relatively low, this should not be a significant factor in most applications mentioned previously.

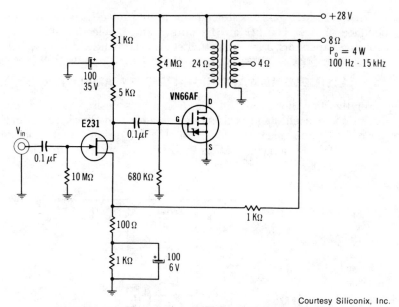

Courtesy Siliconix, Inc.

Fig. 5-1. A 2-transistor, 4-watt VMOSFET audio amplifier.

Design Considerations

The 4-MΩ and 680-kΩ resistors form a voltage divider to bias the gate of the VN66AF VMOSFET at about 4.07 volts, and provide a quiescent operating point for the output transistor of about 0.4 ampere to 0.5 ampere at 28 volts. The power dissipation is, therefore, 11 watts to 14 watts, which will require adequate heat sinking to prevent transistor derating from a rated power dissipation of 15 watts at 25°C.

The 1-kΩ resistor and 100-μF capacitor in the drain circuit of the input transistor provide filtering for the power supply, preventing feedback and 60 Hz from appearing at the output. The 5-kΩ resistor acts as the load for the input transistor and establishes the voltage gain for this stage.

The 1-kΩ resistor and 100-μF capacitor in the source circuit of the input transistor provide dc bias for this transistor without significantly affecting the ac gain, except at low frequencies where the impedance of the capacitor becomes large.

The 100-ohm resistor in series with the input source lead of the transistor and the 1-kΩ resistor from the output of the source lead provides a "voltage-sample voltage sum feedback ratio" of 1:11, which reduces the gain, but also reduces distortion and increases bandwidth.

With a 10-MΩ gate resistor on the input, and a bias circuit impedance of over 500 kΩ on the output stage, selection of 0.1-μF coupling capacitors prevents any effect on the bandwidth.

A SIMPLE 15-WATT HIGH-FIDELITY AMPLIFIER

For the amplifier shown in Fig. 5-2, it was desired to have low distortion in each stage and enough negative feedback to reduce the output impedance to a level suitable for driving quality loudspeakers. In addition, the FET is used because the curvature in the

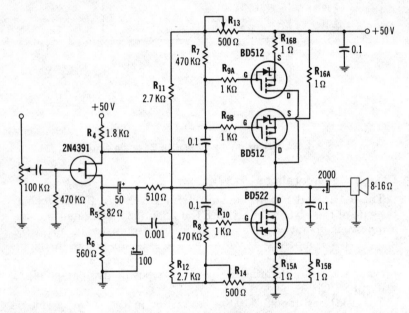

Fig. 5-2. A 15-watt high-fidelity amplifier.

transfer curve is primarily second order. The distortions produced are principally even harmonics and sum and difference frequency components that are less objectionable than odd harmonics, odd-order combination frequencies, and cross-modulation produced by the odd-order curvature typical of bipolar devices.

A complementary pair of VMOS transistors are used for the output stage as their transfer curves are linear over a major portion of their useful current range. A pair of ITT BD 512 p-channel devices in parallel closely match a single BD 522 n-channel transistor (ITT), as n-channel devices of other manufacturers do not closely match the BD 512.

The 2N4391 junction FET is linear above a drain current of 4 mA. Although it is specified for switching applications to achieve high zero-gate-voltage drain current (I_{DSS}), gain, and low small-signal drain-source "on" resistance (r_{ds}), it possesses a linear gain characteristic.

Circuit Operation

The 2N4391 driver can produce a peak-to-peak voltage swing of 20 volts, having a total harmonic distortion of 0.9% with R_6 bypassed, or 0.2% unbypassed. The circuit values were adjusted empirically for maximum output swing. The quiescent drain current is approximately 10 mA.

The complementary output stage operates in the common source configuration to provide voltage gain. Consequently, a bias network is required for each half of the pair. Resistors R_{11}–R_{14} are used to set the quiescent current and the midpoint voltage for the drains. The quiescent current must be greater than 150 mA to achieve acceptable crossover distortion, as graphically shown in Fig. 5-3. Since VMOS devices are temperature stable, little, if any, stabilization is needed. However, the gate threshold voltage characteristic varies several tenths of a volt from device to device, which requires adjustment of R_{13} and R_{14} to achieve required quiescent values.

To avoid designing a complex bias network, ac coupling is used. Resistors R_9 and R_{10} are used to prevent high-frequency parasitic oscillations and excessive loading of the driver stage if zener protected VMOS outputs are used. If a single BD 522 is used for the output stage, then R_{15A} should be placed in parallel with R_{15B}. Although not required to prevent current hogging, source resistors R_{15} and R_{16} are used here to improve current sharing and to reduce distortion in the output stage as a perfect match between the *n*-

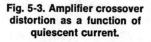

Fig. 5-3. Amplifier crossover distortion as a function of quiescent current.

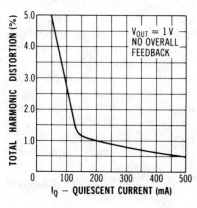

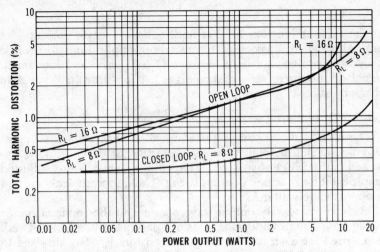

Fig. 5-4. Total harmonic distortion as a function of power output.

channel and p-channel is not possible. Amplifier distortion is shown in Fig. 5-4. Although it could be reduced with additional negative feedback; however, the gain is rather low.

A 40-WATT HIGH-FIDELITY AMPLIFIER*

Fig. 5-5 shows the circuit for a low-distortion (0.04% @ 1 kHz, 40 W) high-fidelity amplifier developed by Siliconix, Inc. The cost is relatively low when VN88AF transistors are used as output devices.

Circuit Operation

Transistors Q_1 and Q_2 operate as a differential amplifier. Q_1 has as its input the input signal of the amplifier, while Q_2 has the feedback of a factor of 1/21 times the output voltage. The output of the Q_1–Q_2 differential pair is fed to another differential pair, Q_3–Q_4, which function as current sources for the current mirrors of IC_1. The output from the current mirrors drives transistors Q_5 and Q_6, which, in turn, drive the output stage. Transistors Q_{11}–Q_{13} are connected as source followers, while Q_8–Q_{10} are connected in a common-source configuration. Local feedback in the form of the R_{14}–R_{15}–C_5 combination enables adjustment of the performance of Q_8–Q_{10} to match that of Q_{11}–Q_{13}. This type of output configuration is known as *quasi-complementary symmetry,* and is often found in audio output stages using junction transistors.

---

* Shaeffer, L., "The VMOS Power FET Audio Amplifier," Design Aid DA76-1, Siliconix, Inc., 1976.

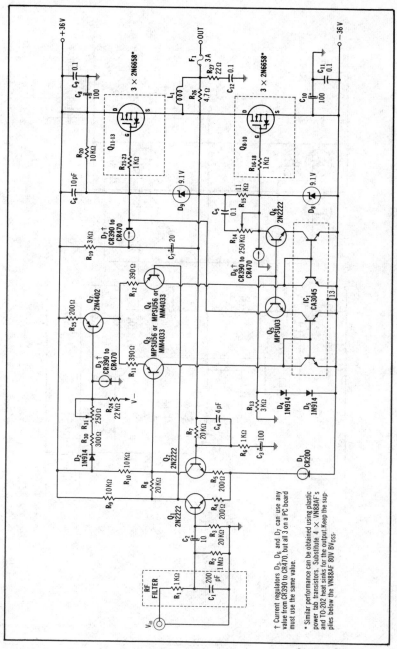

Courtesy Siliconix, Inc.

Fig. 5-5. A 40-watt high-fidelity amplifier.

57

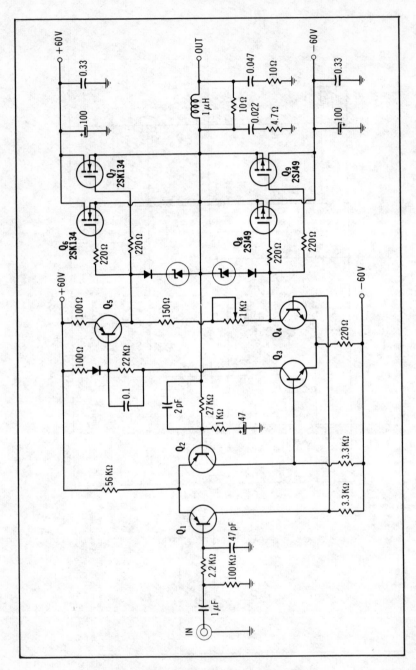

Fig. 5-6. A 100-watt, ultralow distortion amplifier.

Diodes D_8 and D_9 prevent the gate voltages of the output transistors from exceeding 9 volts, thus limiting the output current to less than 2 amperes per transistor. This current limiting, when combined with the output fuse, prevents destruction of the output transistors.

A 100-WATT AUDIO AMPLIFIER

Shown in Fig. 5-6 is the schematic for an ultralow-distortion (less than 0.01% @ 100 W) 100-watt amplifier. While the specified output transistors are not of the typical V-groove construction, they nevertheless are fabricated with short channels so that their characteristics are very nearly similar to those of standard VMOS devices.

CLASS D SWITCHING AUDIO POWER AMPLIFIERS

Class A audio amplifiers have a maximum efficiency of 50%. That is, at least 50% of the energy supplied to the amplifier is lost as heat. While this figure does not seem bad, a 100-watt amplifier must be equipped with sufficient heat sinking in order to dissipate 100 watts of heat. However, these heat sinks could easily cost more than the output transistors.

On the other hand, Class B operation offers an improvement over Class A in that its theoretical maximum efficiency is 78.5%; however, most designs give somewhat less efficiency in order to minimize crossover distortion.

As another alternative, Class D amplification offers a theoretical efficiency of 100% over a wide range of output power levels. Why, then, has it not been used as a practical audio amplifier? The answer is that low-cost transistors capable of operating at high frequencies to minimize distortion have not been available. Furthermore, waveform generator designs have been complex, discouraging attempts to work with Class D amplifier designs.

As shown in Fig. 5-7, the audio signal is continuously compared to the 500-kHz triangle waveform with a comparator. If the audio level is more positive than the triangle waveform level, the output of the comparator is positive, whereas it will be negative for input levels that are less. The output pulse width of the comparator is proportional to the amplitude of the input signal because the peak-to-peak input voltage is always kept at a level that is less than the peak-to-peak voltage of the triangle wave. The resultant output pulses are, in turn, used to switch positive and negative power supplies on and off, consequently dumping more or less current to

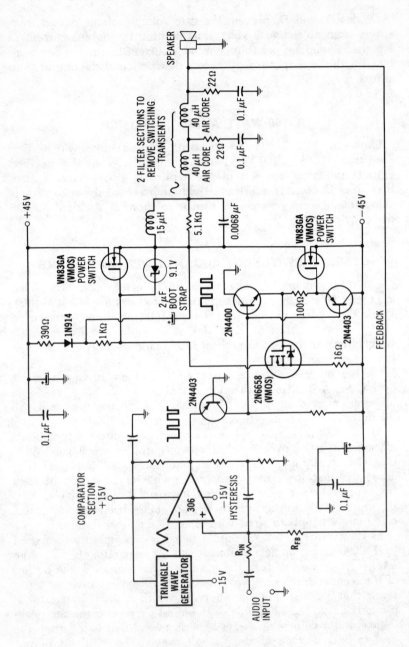

(A) 100-watt Class D switching.

Fig. 5-7. Class D

the load. A low-pass filter in series with the load removes the switching transients and allows reconstruction of the input waveform.

Since the output power devices are either on or off, very low power is dissipated, so that a high efficiency can be achieved. It should be pointed out that this particular circuit is not "ready to go" and connect to a stereo system, for example. Unlike previous circuits discussed in this chapter, the component values shown will, however, enable one to experiment and develop this technique.

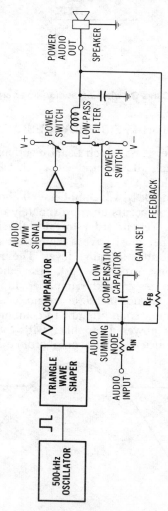

(B) Simplified Class D.

audio power amplifier.

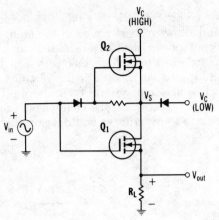

Fig. 5-8. Class G switching audio power amplifier.

ANOTHER HIGH-EFFICIENCY DESIGN: CLASS G

One other amplifier class which bears mentioning is the *Class G* audio amplifier, which allows an efficiency that is greater than the standard Class AB design.

An analysis of the Class AB amplifier will show that the worst-case power dissipation occurs at approximately one-third the maximum power output. Yet, in order to faithfully reproduce musical transients, an amplifier must have a significant amount of reserve power above the normal listening level. These requirements are often weighed against excessive cost, size, and heat dissipation requirements for the design of the amplifier. In effect, the result is a compromise that may not please everyone.

Class G amplification is an attempt to combine the benefits of a low-cost moderately powered amplifier with the capability of reproducing audio transients that would ordinarily be clipped and

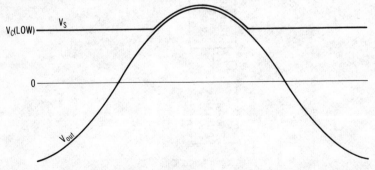

Fig. 5-9. Class G output voltage and source voltage versus transient voltage.

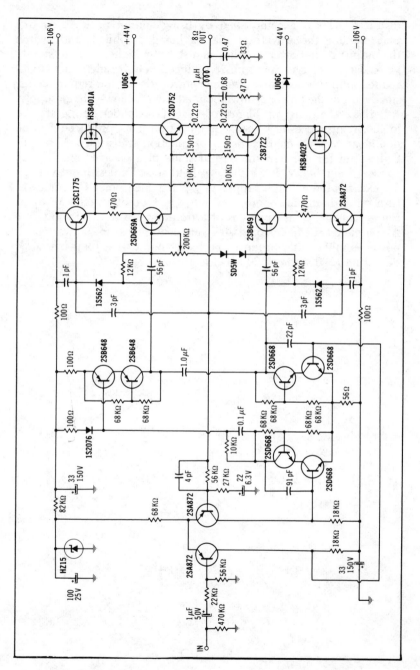

Fig. 5-10. A 200-watt continuous operation Class G amplifier.

distorted. This is accomplished by using audio-type logic circuits that monitor the waveform that is being amplified, and switch the output stage power supplies to a standby higher voltage supply when large voltages must be reproduced. The result is a marked reduction in the heat sink size and power dissipation requirements, since most of the power is now obtained from the low-voltage supply.

Perhaps the major problems for Class G amplifier design are the logic circuits necessary and switching transients produced by the nonlinear turn-on characteristics of junction transistors. A simplified circuit for one-half the output stage of a Class G amplifier is shown in Fig. 5-8. VMOSFET Q_2 is held cut off unless input voltage V_{in} exceeds low-voltage supply $V_{C\,Low}$ at which point Q_2 begins to turn on and supplies more power to Q_1. Output voltage V_{out} and source voltage V_S as functions of a transient are shown in Fig. 5-9. A complete schematic for a 200-watt continuous or 400-watt peak power amplifier that produces no more than 0.01% total harmonic distortion (200 watts @ 20 kHz) is given in Fig. 5-10.

CHAPTER 6

Radio Frequency Applications

INTRODUCTION

In the previous chapter we discussed the difficulty of obtaining linear performance from junction transistors, their susceptibility to thermal runaway and second breakdown. However, in spite of these factors, reasonably reliable and rugged designs for rf amplifiers have been available for some time, using "ballasted emitter" rf transistors.

The ballasted emitter transistor reduces the susceptibility of the junction transistor to thermal runaway by utilizing multiple emitters that have some finite resistance in the emitter region. This finite resistance, in turn, injects a small amount of negative feedback that balances the current and, thus, prevents current hogging and thermal runaway. It does not, however, prevent second breakdown; and it does not significantly increase the normally very low input impedance of the junction transistor at radio frequencies.

Because VMOS power transistors possess very short channels and use majority current carriers, their cutoff frequency is on the order of 600 MHz. In addition, their input impedance at radio frequencies is several tens of ohms, whereas junction transistors have a typical impedance that is approximately a factor of 10 lower. Consequently, VMOS devices are more adaptable to impedance matching.

The lack of minority carrier storage time makes available the use of higher classes of amplification (D, E, and F), which provides increased efficiency. In addition, the input characteristics of the

VMOS do not appreciably change with input power changes. When coupled with a very low noise figure, the possibility of low-level amplification utilizing VMOSFETs becomes attractive.

This chapter discusses several published designs using VMOS-FETs in the design and construction of low-power transmitters and rf power amplifiers.

OBJECTIVES

In this chapter you will learn:

- The advantages of VMOSFET self-protecting characteristics.
- The advantages of VMOSFET high input impedance in rf amplifiers.

A SINGLE VMOS 40- TO 220-MHZ BROADBAND AMPLIFIER

Fig. 6-1 shows the schematic for a 40- to 220-MHz broadband amplifier that is capable of delivering 4.4 watts at approximately 9 dB power gain.[*] Unlike many claims for broadband performance, this amplifier, by virtue of a negative feedback circuit, performs with a flat (±0.5 dB) response over its entire operational bandwidth.

Two interesting features are apparent in the circuit; first, the simple 4:1 transformer T_1 (4 turns No. 22 AWG twisted pair on an Indiana General F625-9Q2 toroid core) for broadband input matching and, second, no output matching circuit. For the equation,

$$R_o = \frac{(V_{DD} - V_{sat})^2}{2P} \qquad (\text{Eq. 6-1})$$

where,

R_o is output impedance,
V_{DD} is drain supply voltage,
V_{sat} is saturated drain-to-source voltage,
P is output power.

With a supply voltage of 25 volts, saturated drain-to-source voltage of 3 volts, and a power output of 4 watts, the output impedance (drain load) is approximately 61 ohms, which is really not that far from the desired 50 ohms, considering there is no matching circuit.

To reach the lower frequency limit of 40 MHz requires a ferrite core with high permeability. In order to reach the 220-MHz upper

---

[*] Taken from the article, "MOSpower FET as a Broadband Amplifier," written by Ed Oxner, *Ham Radio*, 9(12):32-35, December 1976.

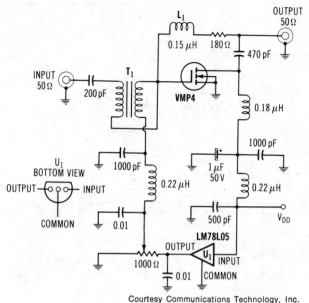

Courtesy Communications Technology, Inc.

Fig. 6-1. Simple 40- to 220-MHz broadband amplifier.

end, 6 to 8 turns of No. 30 AWG enameled wire on a 1-MΩ 1/2-watt resistor are used for inductor L_1.

An interesting aspect of this wideband amplifier is that its performance does not seem to be dependent upon whether it is used for small-signal applications, such as in the front end of a receiver, or for power amplification (1 to 2 watts). For this amplifier, the wideband noise is literally unmeasurably low. The VMP4 *Mospower*® FET has a typical small-signal noise figure of 2.4 dB at 146 MHz with a properly matched input circuit. However, it should be noted that this particular circuit uses a 4:1 transformer and is not properly matched for optimum results. In addition, the VMP4 VMOSFET is not sensitive to load mismatches.

A 1.8- TO 54-MHZ 16-WATT BROADBAND AMPLIFIER

The circuit for a 16-watt broadband amplifier, which may be used either in Class B or Class D using a pair of VMP4 VMOS power FETs, is shown in Fig. 6-2.* Operating in a push-pull fashion,

---

® Mospower is a registered trademark of Siliconix, Inc.

* From the article "MOSFET Power Amplifier for Operation from 160-6 Meters," written by Frederick H. Raab, *Ham Radio*, 11(11):12-17, November 1978.

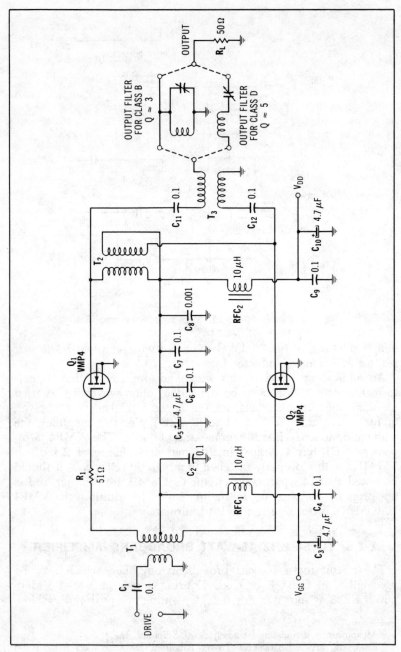

Courtesy Communications Technology, Inc.

Fig. 6-2. Simple 1.8- to 54-MHz, 16-watt broadband amplifier.

this circuit may be operated in either Class B for *linear* power amplification or in Class D for *highly efficient* power amplification by use of the proper output filter.

For Class B operation, the output filter bypasses any harmonic components in the output current, allowing only the fundamental, or *carrier*, frequency to reach the load. The sinusoidal voltage generated there becomes the drain voltage waveform with the addition of the supply voltage. The output power is then expressed as

$$P_o = \frac{V_{DD}^2 R}{2(R + r_{on})^2} \qquad \text{(Eq. 6-2)}$$

where R is the drain load resistance. At peak power output, the efficiency of the amplifier is given by

$$\eta \simeq \frac{\pi R}{4(R + r_{on})} \leq 78.5\% \qquad \text{(Eq. 6-3)}$$

Class D operation also uses a pair of devices in push-pull, but they are controlled by the driving signal to act as switches. The output filter now passes only at the fundamental, or *switching*, frequency, so that the sinusoidal output voltage is equal to the fundamental frequency component of the square wave on the secondary winding of the transformer. The output power is expressed as

$$P_o = \frac{8R V_{DD}^2}{\pi^2 (R + r_{on})^2} \qquad \text{(Eq. 6-4)}$$

Although the Class D amplifier can ideally achieve an efficiency of 100%, the efficiency is approximated by

$$\eta \simeq \frac{R}{R + r_{on}} \qquad \text{(Eq. 6-5)}$$

at most output levels.

Output Design

The VMP4 allows a maximum drain current of 1.6 amperes and a maximum drain voltage of 60 volts, which is made up of a +30-volt ac variation around a supply voltage of +30 volts. If the saturation voltage is zero, power outputs are 24 watts and 30.4 watts for Class B and Class D, respectively. For these outputs, load resistances of 30 W/1.6 A = 18.75 Ω (Class B) and $(4/\pi)(30$ W/1.6 A) = 23.87 Ω (Class D) are required. The load resistance required to obtain the maximum output power when the saturation "on" resistance is nonzero is determined by subtracting the saturation resistance from the load resistances for zero saturation resistance. Therefore, assuming a typical saturation resistance of 2.6 ohms, the maximum power loadlines are $18.75 - 2.6 = 16.15$ Ω and $23.87 - 2.6 = 21.27$ Ω for Class B and Class D operation, respectively.

Optimum broadband performance is accomplished by the use of "transmission-line" transformers, which limit the choice of impedance transformations. A convenient transformation is a ratio of 4:1, with a 50-ohm load; the load line is then 12.5 ohms. The hybrid transformer, T_2, is constructed by winding two turns of 25-ohm transmission line through two parallel stacks (2.86 cm long) of Ceramic Magnetics model CN-20 ferrite. The 25-ohm transmission line is obtained by connecting two 50-ohm miniature coaxial cables in parallel.

Balun T_3 requires a 50-ohm characteristic impedance line, and is constructed by winding two turns of miniature coaxial cable through two parallel stacks of CN-20 ferrite. The combination of T_2 and T_3 provides a load that is primarily resistive over the frequency range when connected to a 50-ohm load.

With a load line of 12.5, the output power is limited by the maximum drain current I_Dmax, so that

$$P_omax = \frac{I_Dmax^2R}{2} \qquad \text{(Eq. 6-6)}$$

$$= \frac{(1.6 \text{ A})^2(12.5 \text{ Ω})}{2}$$

$$= 16 \text{ watts}$$

The 1.6-ampere peak drain current then requires an average, or dc, supply current of $(1.6)(2/\pi) = 1.02$ A. For Class B operation, the drain voltage swing is $(1.6 \text{ A})(12.5 \text{ Ω}) = 20$ volts at peak output. To avoid saturation, the drain voltage must be greater than $(1.6 \text{ A})(2.6 \text{ Ω}) = 4.2$ volts. However, the supply voltage must then be equal to 24.2 volts for maximum power output, but the efficiency is reduced.

For Class D operation, the 4.2-volt drop across the saturation resistance of the VMOSFET opposes the fundamental frequency component of the ideal square wave, requiring a supply voltage of $(24.2)(\pi/4) = 19.0$ volts for maximum output. Consequently, the efficiency for Class D is 82.8%, while it is 65% for Class B.

Input Design

For a peak drain current of 2.6 amperes, the bias voltage, as determined from the VMP4 data sheet, should be about 2 volts with a peak sinusoidal voltage of about 6 volts. A 4:1 voltage reduction, however, produces a more convenient output for the driver. Since the input impedance of the VMP4 can vary from 2 kΩ at 1.8 MHz to 59 ohms at 54 MHz, precise broadband matching is very difficult.

If it can be assumed that the input drive power is relatively small in comparison to the output power, matching of the input to the power amplifier is not critical for most applications.

Transformer T_1 is a single-turn primary with an 8-turn center-tapped secondary wound on a 1-cm diameter Ferroxcube 3B7 core.

Additional circuits and design techniques for VMOSFET rf amplifiers can be found in the following three articles:

1. Oxner, E., "Build a Broadband Ultralinear VMOS Amplifier," *QST*, 63 (5):23-26, May 1979.
2. Hayward, W., "A VMOS FET transmitter for 10-Meter CW," *QST*, 63 (5):27-30, May 1979.
3. Frey, G., "VMOS Power Amplifiers," *EDN*, 22(16):83-5, September 5, 1977.

CHAPTER 7

Power Supply Applications

OBJECTIVES

In this chapter you will learn:

- The basic operation of switching regulators, and the advantages of using VMOSFET switching elements.
- The use of VMOSFETs in inverter circuits.
- The simplicity of using VMOSFETs as linear pass elements.
- The fast protection that can be afforded using VMOSFETs as overvoltage protection.

SWITCHING REGULATORS

The linear integrated circuit regulator has been in use for several years, and serves as the heart of the power supplies for a majority of the electronic devices on the market. A major disadvantage of the linear regulators is their low efficiency and the need for transformed voltages, requiring very large input capacitors.

For example, the popular LM309K is capable of supplying 5 volts ($\pm5\%$) at 1 ampere. The minimum input voltage required is 7 volts, and since the input voltage drops continually during most of an ac half-cycle, the ac input to the rectifier must be greater than 7 volts. Therefore, the input to the rectifier is typically 10 to 12 volts ac. For an average input voltage to the regulator of 10 volts, the efficiency is, therefore, only 50%. One-half of the input power is wasted on the regulator itself.

On the other hand, the switching regulator takes advantage of the physical inductance and capacitance to maintain a regulated output within a specified tolerance. As shown in the basic switching regulator circuit of Fig. 7-1, when the output voltage is less than the reference voltage, switch S_1 is closed. The current through the inductor is now allowed to increase, and continues until the output voltage rises to the desired output level. Consequently, the current through the inductor is now greater than the output current, since it is supplying current to both filter capacitor C_1 and the load.

Now the comparator turns the switch off, preventing any further increase in the current to the inductor. Diode D_1 now conducts the inductor current which, in turn, allows the output voltage to increase further until the load current is greater than the current through the inductor.

The capacitor now begins discharging, and supplies current to the load until the output voltage drops below the reference voltage, and the process is again repeated. If the switch is ideal, no power losses occur in the circuit, except at the load.

However, there are some disadvantages in using switching regulators, which require the use of inductors and capacitors and are somewhat more complex than linear regulators. In addition, the operation of the switching regulator requires some hysteresis in order for switching to occur, and there is inherently a small, but finite, amount of ripple voltage present on the output voltage.

In order to minimize the size of the inductor and capacitor, the switching frequency must be increased, which is not always possible. To ensure efficient operation, the pass transistor used as the switch must be driven into full saturation. However, this requires significant base drive power and, in turn, reduces the maximum allowable frequency due to the excess minority carrier storage in the base region of the transistor. In addition, switching losses increase rapidly with frequency. Therefore, if the switching frequency is increased too high, saturation does not occur, and power losses are further increased.

When VMOSFETs are used as pass transistors, the switching drive power is drastically reduced, since we need now only to either

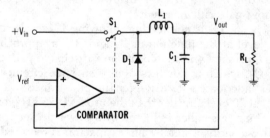

Fig. 7-1. Basic switching regulator.

charge or discharge the very small input gate capacitance of the device. Consequently, the maximum switching frequency is substantially higher than for bipolar junction transistors, and reduces the sizes for the inductor and capacitor.

The use of VMOSFETs, however, may entail a reduction in the efficiency at low frequencies, as compared to junction transistors. This is due to the relatively high "on" resistance of VMOSFETs, which can be as high as 3 ohms for 2-ampere devices, while the resistances of junction transistors are an order of magnitude lower. However, there are some VMOS devices that are comparable to the "on" resistance of the junction transistor, such as the IRF 100 (International Rectifier), which is capable of switching 16 amperes at 80 volts.

The reliability of VMOSFETs and their inherent thermal stability have induced some manufacturers of switching regulators to use VMOS designs. The slight increase in cost is far outweighed by the reduction in size and increased reliability.

As an example of VMOS switching regulators, Fig. 7-2 shows the schematic for a 10-ampere 5-volt regulator operating at a maximum frequency of 200 kHz.

This circuit contains a "soft-startup" feature, made up of C_{13}, R_{12}, R_{13}, D_6, D_7, and Q_3, which limits the drive pulse amplitude to Q_1 at an exponentially increasing value when the circuit is first energized. Consequently, this prevents an excessively large inductor current when the 710 comparator first switches the drive pulse off. It should be recalled that the output voltage of the switching regulator continues to rise after the pass transistor turns off since the inductor current continues to charge the output capacitor. The soft-startup circuit, therefore, reduces the rate of rise of inductor current when the power is first turned on, and prevents a large overshoot in output voltage. Beyond this, the startup circuit has no further effect on the operation of the regulator.

To ensure good efficiency, the gate-source voltage must rise as rapidly as possible during normal circuit operation. A "bootstrap" circuit, consisting of R_5 and C_3, provides a 43-volt drive to the gate to ensure rapid gate enhancement.

When Q_1 is off, D_5 is conducting, allowing C_3 to charge toward the voltage level at the junction of R_4 and R_5, or approximately 19 volts. When Q_1 is on, Q_2 is off and the only discharge path is through the gate of Q_1, which is now essentially an open circuit. In effect, this provides adequate gate drive for the pass transistor.

Operation at 200 kHz, rather than the usual 20–25 kHz, has several advantages. First, a smaller inductor with lower dc (copper) losses is permitted. Secondly, a smaller filter capacitor is required. Thirdly, the regulator responds faster to sudden changes in the load.

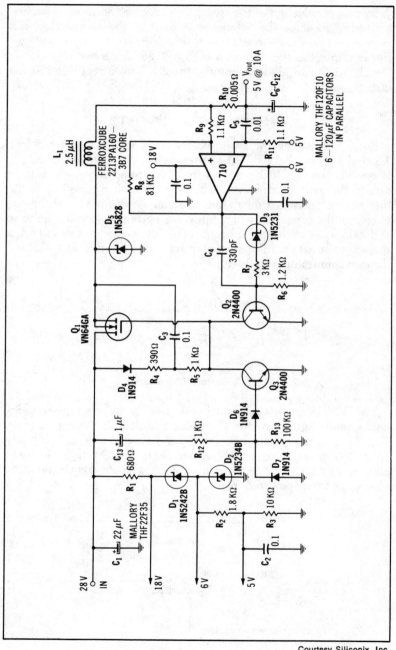

Courtesy Siliconix, Inc.

Fig. 7-2. A 10-A, 5-V (200 kHz) regulator.

75

By proper selection of the switching inductor and input capacitor, and adjustment of the reference voltage of the comparator, this regulator circuit may be adapted to a wide range of output voltages. For example, the selection of a high-voltage pass transistor, such as the IRF 300, can allow operation directly from a rectified 110-volt line.

VMOS INVERTER CIRCUITS

Fig. 7-3 illustrates the basic scheme for a VMOS inverter circuit. The inductor is "charged" with current when the switching transistor is on. No power is delivered to the load from the supply voltage during this period. However, when the switching transistor is off, the current flowing through the inductor flows through the diode to charge the capacitor, which filters the current pulses and helps to maintain a constant load current. (One drawback of this circuit is the requirement that the switching transistor be able to withstand the maximum output voltage.)

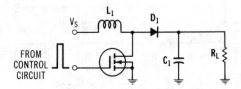

Fig. 7-3. Basic VMOS inverter.

The VMOS inverter circuitry can readily incorporate complex analog and digital functions without additional interface circuits. For example, the circuit shown in Fig. 7-4 may be used as a digitally controlled inverter to supply either ac or high-voltage dc upon command, such as from a microcomputer, remote switch, or other sources of lower-powered logic.

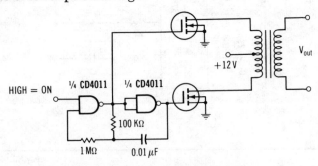

Fig. 7-4. Digitally controlled inverter.

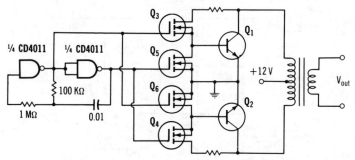

Fig. 7-5. Hybrid VMOSFET/junction-transistor inverter.

While the circuit shown is not highly efficient with the 2-ampere VMOSFETs that are presently available, it is nevertheless quite simple. The typical 10-ohm "on" resistance would result in a maximum efficiency of approximately 68%. However, a higher supply voltage will result in higher efficiency.

A more efficient low-voltage inverter can be constructed by using a hybrid VMOSFET/junction-transistor arrangement like that shown in Fig. 7-5. Transistors Q_1 and Q_2 serve as switching transistors for the inverter, while VMOSFETs Q_3 and Q_4 are both drivers for Q_1 and Q_2 and serve to interface them to the CMOS logic. Thus, Q_5 and Q_6 allow for high-frequency operation by quickly discharging the base capacitance of the "off" transistor, thus preventing both transistors from being "on" simultaneously. With this arrangement, the efficiency can easily be greater than 90%, and low-cost transformers may be used in lieu of expensive multiple winding inverter transformers.

LINEAR REGULATORS

For the circuit designer whose designs frequently require the use of linear regulator circuits, VMOS devices can offer several distinct circuit simplifications. The power gain of the VMOSFET ($10^6 - 10^7$) allows for the use of very low power regulator circuits, while its infinite current gain eliminates the need for driver transistors in the final circuit. Any need for increased current is met by simply placing additional VMOS devices in parallel and adjusting the size of the current sense resistors.

Fig. 7-6 illustrates the circuit for a simple linear regulator capable of supplying 5 volts at 2 amperes, with no additional current limiting circuitry. Since a 2-ampere load will require $(15 - 5)(2) = 20$ watts of power dissipation by the pass transistor, adequate heat sinking is required. However, the use of two power-tab devices operating in parallel should be adequate if the 10-volt zener diode is reduced

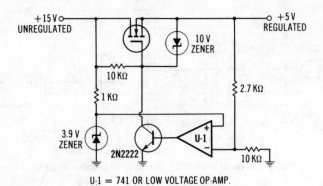

Fig. 7-6. Linear regulator circuit.

to approximately 6.2 volts in order to limit the output current at 2 amperes.

OVERLOAD PROTECTION

Since the current through a VMOSFET is limited by the gate-source voltage, overload protection becomes extremely simple. By placing a zener diode between the gate and source leads, the output current can then be limited to the desired value. For example, 10

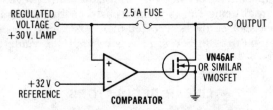

Fig. 7-7. Overvoltage protection circuit.

volts of enhancement will provide about 10 amperes to 12 amperes of current through a VN84GA VMOSFET. By inserting a 7.1-volt zener diode between the gate and source, the output current is now limited to approximately 6 amperes. With adequate heat sinking, the device can operate with a short circuit continuously without damage.

Overvoltage protection can be provided faster than with the present "crowbar" techniques, as shown by the circuit in Fig. 7-7. When the output of the regulator increases to 32 volts, the output of the comparator will be positive, turning the VMOSFET on, which will blow the fuse. While fuses are not free, they nevertheless are much less expensive than the devices they protect.

CHAPTER 8

Microcomputer Applications

INTRODUCTION

Microprocessors have produced the capability of designing digital logic circuits with the ability to perform complex decision-making tasks. The large-scale integration techniques that are used to manufacture microprocessors and their related support circuits do not readily lend themselves to handling the power necessary to control "real-world" devices. Unfortunately, this leaves one with the task of interfacing the low-powered logic signals to power-hungry displays, or controlled devices.

VMOSFETs, with their infinite current gain, are an excellent choice as an interface device. This chapter discusses how low-powered logic can be interfaced to control several amperes of current, or hundreds of volts of input with a minimum number of components.

OBJECTIVES

In this chapter you will learn:

- The ease with which microwatt logic signals can be interfaced to high-power devices using VMOSFETS.
- The techniques of interfacing various logic families to VMOS-FETs.
- The advantages of VMOSFET interface circuits.

LOGIC-LEVEL INTERFACING

Interface requirements of the VMOS are greatly simplified by their very high input impedance and their small, but significant, threshold voltage.

Consider, for example, the CMOS-to-VMOS interface shown in Fig. 8-1. For a 10-volt VMOS logic supply, the circuit shown is capable of switching 2 amperes at 80 volts in 60 ns, which is well within the requirements of most microprocessor applications.

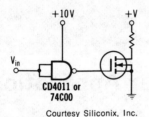

Fig. 8-1. Driving VMOS with a CMOS gate.

Courtesy Siliconix, Inc.

A ready application of the simple interface circuit could be the interfacing of a dot-matrix printer head with a character generator, as diagrammed in Fig. 8-2. By selection of the appropriate logic supply voltage, the current to the dot solenoids may be programmed to the desired value. This application results in a factor of 10 decrease of discrete components compared to typical print solenoid drivers.

TTL Interfaces

Many interface applications, however, do not have voltage levels that are readily compatible with the driving levels required for VMOSFETs. For example, the current levels available with 5-volt TTL signals applied to the gate are adequate when a single resistor is used, as shown in Fig. 8-3. The 10-kΩ resistor provides a pull-up to the full 5-volt logic level, since most TTL logic 1 outputs are at least one diode drop (approximately 0.7 volt) below the supply voltage. The 0.3-volt logic 0 output level is well below the gate threshold voltage for VMOSFETs, thus ensuring that the load current will be zero for a logic 0 output.

For TTL applications where the full drain current capacity is required, open collector output devices may be used, as shown in Fig. 8-4. While this arrangement may be suitable for many applications, the turn-on time may not be adequate for high-speed situations. By replacing the 10-kΩ pull-up resistor with an emitter follower, shown in Fig. 8-5, the switching speed can be markedly increased.

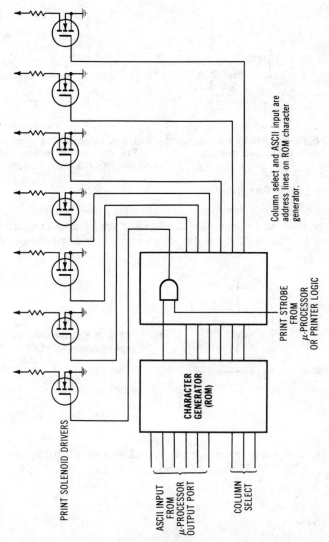

Fig. 8-2. Interfacing a dot-matrix head with an integrated circuit character generator.

A more universal interface for TTL and VMOS devices utilizes an MH0026 clock driver, which is designed to translate the lower level TTL logic signal to higher level (15-volt) VMOS with sufficient current capacity to drive high capacitance loads (Fig. 8-6). Switching times using this arrangement are on the order of 30 ns with a rise time of less than 10 ns.

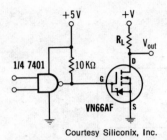

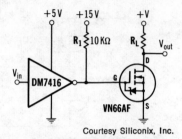

Fig. 8-3. Driving VMOS with standard TTL integrated circuits.

Fig. 8-4. Using a 7416 open-collector TTL device with pull-up resistor for full drain current.

Other Logic Families

For other logic families, a suitable interface is possible using a comparator, or an operational amplifier wired as a comparator, as shown in Fig. 8-7. By suitable selection of the reference voltage

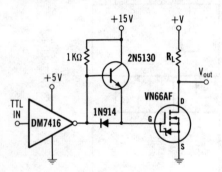

Fig. 8-5. Increasing switching speed using an emitter follower.

Courtesy Siliconix, Inc.

(V_{ref}), the comparator circuit can interface virtually any logic family with VMOS devices, while producing switching times that are less than 40 ns. For positive logic families, the reference level

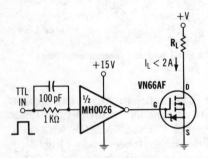

Fig. 8-6. Voltage translation using an MH0026 driver.

Courtesy Siliconix, Inc.

should be greater than the maximum logic 0 voltage plus the noise margin. For TTL, or other 5-volt logic families, this is about 1 volt, while it should be about −1.2 volts for ECL levels.

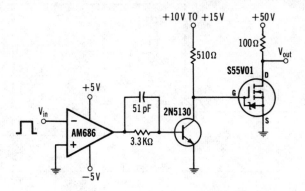

Courtesy Siliconix, Inc.

Fig. 8-7. Op-amp comparator, high current interface.

Since ECL devices are high speed, the comparator circuit of Fig. 8-7 is not able to handle ECL speeds in most applications, but should only be used for those situations where the input pulses are relatively long. Fig. 8-8 shows an interface circuit that is suitable for interfacing high-speed pulses from ECL devices with a VMOS device.

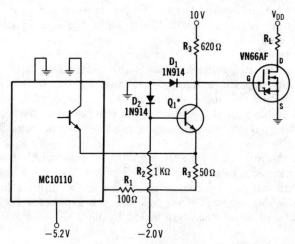

*Q_1 — USE A HIGH f_t RF BIPOLAR TRANSISTOR

Courtesy Siliconix, Inc.

Fig. 8-8. ECL-to-VMOS interface.

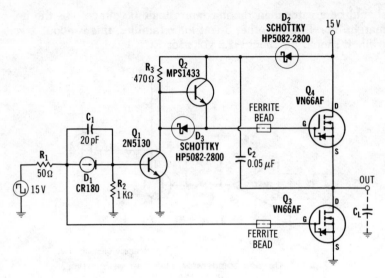

Fig. 8-9. VMOS high-speed line driver.

LINE DRIVER CIRCUIT

Perhaps one of the more difficult design problems to appreciate, until one attempts it, is transmitting logic level signals over long distances. Logic signals contain high-frequency components that can easily be reflected back and forth along the signal path, which results in data "glitches."

Also, these long lines may act as an antenna and pick up electrical noise signals, especially when the termination is of low impedance. As a possible solution, one may choose to use shielded cable to reduce the noise pickup, but the majority of the shielded cables available exhibit low characteristic impedances that can cause increased signal reflection unless terminated by the same low impedance.

One technique that reduces signal reflection while permitting high data transfer rates along shielded cables is to use a line driver circuit that can rapidly charge or discharge the input capacitance of the cable. Shown in Fig. 8-9, this circuit is capable of baud times on the order of 50 ns when terminated by 1000 pF.

PERIPHERAL INTERFACING

Considering the inherent stability and self-protecting features of VMOSFETs, it is difficult to make errors in using them as peripheral drivers. In this section, several circuits are discussed that demonstrate the simplicity of peripheral control with VMOS devices. If load currents of 400 mA or less are required, then +5-volt logic levels should furnish enough gate drive. For higher load currents,

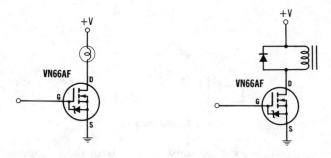

Fig. 8-10. A simple lamp driver.   **Fig. 8-11. Relay or solenoid driver.**

logic-level interface circuits may be used, or VMOSMETs may be connected directly in parallel.

A simple lamp driver is shown in Fig. 8-10. The only precaution is that the maximum drain-source voltage rating must be greater than the supply voltage. Fig. 8-11 is used for driving relays or

Fig. 8-12. Magnetic core driver.

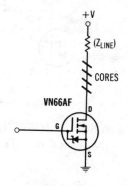

solenoids. A series resistor which can be used to increase the speed of the circuit by reducing the RL time constant at the cost of increased power loss is optional, but a higher supply voltage is required.

For driving magnetic cores, the circuit shown in Fig. 8-12 can be used, in which case the input voltage must be sufficient to ensure that the resultant drain current will produce the necessary saturation flux in the driver cores. On the other hand, a single VMOSFET can be used to control an ac load with the circuit of Fig. 8-13. The ac load current must be less than the maximum drain current rating,

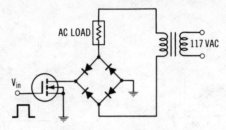

Fig. 8-13. Ac load control.

while the peak ac voltage must be less than the maximum drain-source voltage rating. With the proper VMOSFET and bridge rectifiers, this circuit can control a 220-volt ac 4-ampere load with only a few microwatts of input signal power! Similar to the lamp driver

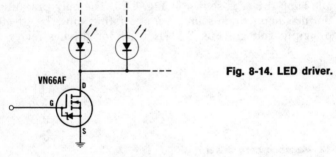

Fig. 8-14. LED driver.

Courtesy Siliconix, Inc.

of Fig. 8-10, VMOSFETs are used to turn LEDs on and off, one or more connected in parallel, as shown in Fig. 8-14.

AN EXAMPLE

As a greater number of surplus computer peripheral devices become available and designs for new devices are conceived, the application of state-of-the-art technology can produce extremely reliable and simple circuitry.

As an example, consider the possibility of an optical paper tape reader utilizing VMOSFET devices. The tape reader basically consists of a reluctance-type stepping motor, a microswitch to sense the

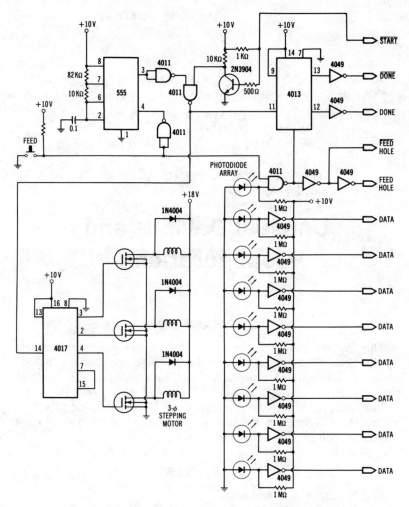

Fig. 8-15. Paper tape reader and stepping motor drive.

end of the paper tape, a light source, and a photodiode array. Fig.
8-15 shows the schematic for the optical paper tape reader with the
I/O signals required for computer interfacing, permitting rates
greater than 120 words per second.

CHAPTER 9

Unusual Devices and Configurations

INTRODUCTION

The bipolar junction transistor has found numerous unusual applications other than as an amplifier. These applications include zener diodes, current sources, current mirrors, rectifiers, variable resistors, etc.

This chapter discusses a number of VMOSFET circuits that utilize these devices in unusual ways, or applications that do not readily fit into previous chapters.

OBJECTIVES

In this chapter you will learn:

- To use a VMOSFET as a high-power constant current source.
- To use a VMOSFET as a variable resistor.
- To use a VMOSFET as an efficient power controller.
- To use a VMOSFET as an analog switch.
- To improve present automotive ignition designs using VMOS-FETs.

A VMOS CONSTANT CURRENT SOURCE

Unlike the bipolar transistor, the VMOSFET constant current source, shown in Fig. 9-1, is relatively insensitive to temperature changes and can supply a wide range of output currents, I_L, so that

$$I_L = \frac{V_{BE}}{R_S} \qquad \text{(Eq. 9-1)}$$

where,

V_{BE} is the base-emitter forward drop of the junction transistor,
R_S is the source resistance.

The base-to-emitter forward drop of the junction transistor is typically 0.7 volt for silicon types. For optimum thermal stability, the collector current I_C is made approximately $I_L/30$.

As an example, consider a constant current source that is to deliver 1 ampere to a load with a nominal 20-volt supply. Then from Equation 9-1,

$$R_S = \frac{0.7 \text{ V}}{1 \text{ A}}$$
$$= 0.7 \text{ ohm}$$

For a 1-ampere drain current, the 2N6656 VMOSFET, for example, requires that the gate-source voltage V_{GS} be approximately 6.2 volts. If the load is short-circuited, the voltage drop across collector resistor R_C is then $20 - 6.2$, or 13.8 volts. The current through R_C for optimum thermal stability is to be approximately $I_L/30$, so that $I_C = 0.033$ ampere. Therefore,

$$R_C = \frac{13.8 \text{ V}}{0.033 \text{ A}}$$
$$= 418 \text{ ohms}$$

in which case a 470-ohm resistor will be sufficient. It should be realized that the VMOSFET must dissipate 20 watts. For drain currents greater than 2 amperes, simply place additional devices in parallel until the required current capacity is achieved.

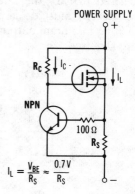

Fig. 9-1. Constant current source.

89

HIGH CURRENT VARIABLE RESISTOR

When a VMOSFET is operated such that the drain-source voltage is approximately 3 volts (i.e., the nonsaturation region), the device exhibits a fairly linear inverse relationship between drain-source resistance and gate-source voltage. For the 2N6656, for example, its gate-source resistance can vary from about 2 ohms ($V_{GS} = 10$ volts) to essentially infinity ($V_{GS} = 0$).

The circuit model for this resistive, or triode, operation is shown in Fig. 9-2 for an *n*-channel device. Note that the diode is reverse biased for normal supply connections, but results in a short circuit for reverse voltage connections. Since this phenomena applies only for low drain-source voltages, "power-tab" devices may be readily substituted for high-wattage low-ohmic resistors at relatively low cost.

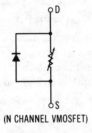

Fig. 9-2. VMOS high current variable resistor model.

EFFICIENT POWER CONTROLS

One simple means for providing variable power to a load is to insert a variable resistor in series with the load. However, this method is very inefficient. As an example, consider the variable light intensity circuit of Fig. 9-3. If a 25-watt bulb is used and it is desired to vary the intensity from one-fourth to full power, then resistor R_1 must be 25 ohms. Consequently, the efficiency is only 50% at minimum output power and decreases if lower output levels are required.

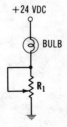

Fig. 9-3. Basic variable light intensity control.

A more efficient approach is obtained by using a constant frequency, variable duty cycle multivibrator driver, like the CMOS circuit shown in Fig. 9-4. In effect, this arrangement allows an extremely wide range of output power levels at constant efficiency, which is determined by the VMOS "on" channel resistance and the resistance of the bulb. For this circuit, the VMOSFET is either

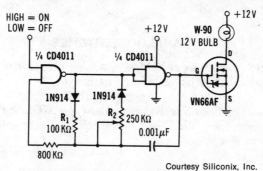

Courtesy Siliconix, Inc.

Fig. 9-4. Variable light intensity control using a variable duty cycle astable multivibrator.

off or on, with the duty cycle of the multivibrator determined by the ratio of R_1/R_2. The type 4011 NAND gate circuit oscillates at about 2 kHz with the components shown so that the light appears to be lit continuously.

The circuit of Fig. 9-4 may also be used to control power to a motor by simply replacing the bulb with a dc motor. As in previous cases, if the current demand is greater than the capacity of the VMOSFET, additional devices may be connected in parallel.

A FAST PULSER CIRCUIT

There exists a variety of applications for which a large current pulse of short duration is required. Prior to the availability of VMOSFETs, either vacuum tube or avalanche transistor circuits were required for rise times of 4 ns to 10 ns.

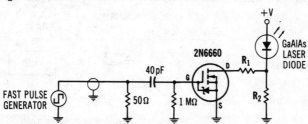

Fig. 9-5. Fast pulse switching circuit for driving a GaAlAs laser diode.

Fig. 9-5 shows a fast pulse switching circuit for driving a GaAlAs laser diode using a 2N6660 VMOSFET, assuming that a fast pulse generator source having a 50-ohm output impedance is used. Resistors R_1 and R_2 are selected as needed to both limit the peak current pulse and maintain the current for the laser diode. With the components shown, current pulses up to 3 amperes are possible from a TO-5 packaged device.

ANALOG SWITCHES

Although electronic switching of analog signals has been successfully accomplished using bipolar transistors, it is not widely done because of the complex circuitry required in addition to inherent distortion. On the other hand, planar-type MOSFETs and CMOS devices have been used, but they both are incapable of handling high power levels.

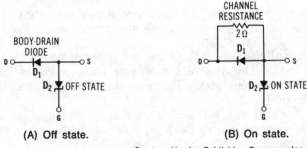

(A) Off state.   (B) On state.

Courtesy Hayden Publishing Company, Inc.
Fig. 9-6. VMOS transistor models.

VMOSFETs, however, provide a reliable method of switching analog signals in audio as well as radio-frequency ranges. Input power requirements are negligible, while power handling capability can be as high as 80 watts.

Fig. 9-6 shows the simplified models of the "off" and "on" states for a VMOS transistor. Diode D_1 is the body-to-drain junction diode,

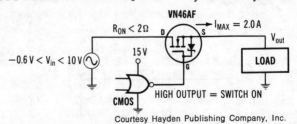

Courtesy Hayden Publishing Company, Inc.
Fig. 9-7. Basic switching circuit limited to positive voltage levels.

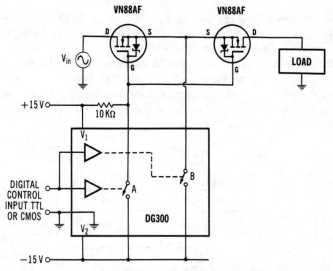

Fig. 9-8. Increasing the dynamic range by connecting two VMOSFETs in series.

while D_2 is the zener gate protection diode. When the device is "on," there is typically a 2-ohm "on" resistance, which is in parallel with D_1.

Unlike CMOS analog switches, the input signal, using the basic circuit of Fig. 9-7, is restricted to *positive* voltages, otherwise the "off" state isolation will be impaired. However, the dynamic range of a VMOS switch can be increased by connecting two devices in

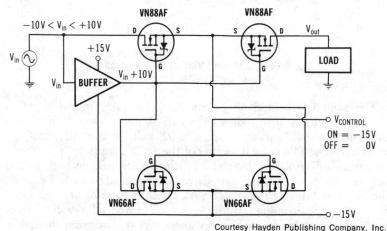

Fig. 9-9. Achieving a constant on-state resistance.

series, as shown in Fig. 9-8. In the "on" state, both halves of the DG300 analog switch are open, so that the gates of both VMOS transistors are "pulled up" to the +15-volt supply through the 10-kΩ resistor. The "on" resistance of this configuration is twice as high as the drain-source resistance of a single transistor, since two transistors are now connected in series. The maximum current that this two-transistor switch can handle is the same as that for the single transistor switch of Fig. 9-7.

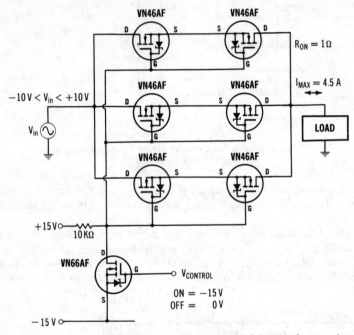

Fig. 9-10. Connecting VMOSFETs in parallel to reduce the on-state resistance.

The switch is turned off by shorting the gates to the negative supply, which, in turn, reduces the gate-source voltage to less than the gate threshold voltage, typically 0.8 volt for two VN88AF devices in series. The second section of the DG300 adds approximately 30 dB of "off" isolation by shunting the signal-leakage path to the negative supply. Since the gate drive is referenced to a fixed supply, the "on" resistance of the switch varies with the analog voltage. Consequently, this variation introduces distortion when the switch drives low impedance loads, such as speakers and transmission lines. For the circuit of Fig. 9-8, the "on" resistance can range from approxi-

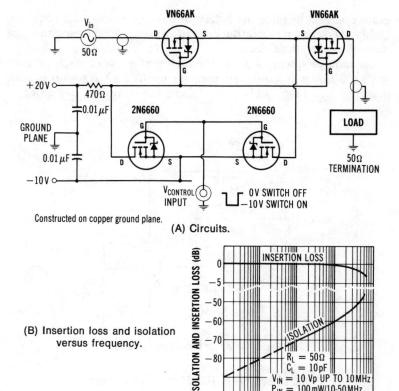

(A) Circuits.

(B) Insertion loss and isolation versus frequency.

Courtesy Hayden Publishing Company, Inc.

Fig. 9-11. High-power, radio frequency switch.

mately 2 ohms to 4 ohms when the input voltage is varied from −15 volts to +11 volts. Fig. 9-9 shows a circuit that is used to achieve a constant "on" resistance of less than 4 ohms over the same input range. In the "on" state, a bootstrap voltage, which tracks the input, drives the gates of the VMOS devices, and thereby keeps the gate-source voltage constant and independent of the input signal. Consequently, any changes in the input level do not modulate the "on" resistance of the switch.

Although the dynamic range of the switch is increased by operating VMOS devices in series, the "on" resistance is increased. However, by operating devices in parallel, as shown in Fig. 9-10, the total "on" resistance is lower than for a single device switch, which is typically 1 ohm for the devices shown. There are three parallel paths, each with two devices in series. Since VMOS devices do not exhibit current hogging, no ballast or balance resistors are required. Unlike the series method, the current capacity is increased to three times that of a single device.

For radio-frequency signals, the high-power switch circuit shown in Fig. 9-11A will handle frequencies up to 50 MHz with a 1-dB insertion loss (Fig. 9-11B). Switching times are on the order of 50 ns.

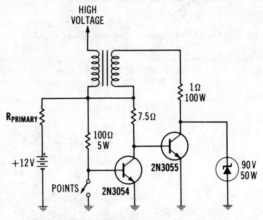

Fig. 9-12. Basic transistor ignition.

AUTOMOTIVE IGNITION SYSTEMS

In much the same way as mechanical switches have been a problem in conventional electronic circuits, they also have been a major source of maintenance problems in standard automotive ignition

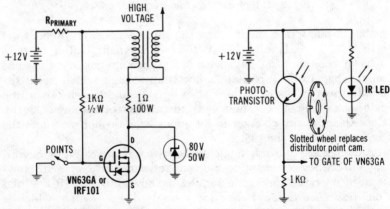

Fig. 9-13. VMOSFET ignition that reduces point current.

Fig. 9-14. Optically coupled ignition circuit.

systems. Recently, automobile manufacturers have moved toward electronic switching to eliminate the high distributor-point current required in the standard Kettering ignition system.

While some designs have eliminated the points entirely, others reduce the point current by using the basic transistor ignition circuit shown in Fig. 9-12. As an alternative, an equivalent circuit using a VMOS transistor is shown in Fig. 9-13, which now eliminates the predriver transistor, reduces point current, and reduces the number of circuit components.

While the circuit of Fig. 9-13 reduces the point current to approximately 12 mA, the distributor points are still subject to fouling as a result of dirt or oil accumulation. However, the points themselves may be eliminated with the circuit of Fig. 9-14. The slotted wheel replacing the points interrupts a light beam from a light-emitting diode to a phototransistor. The very high gain of the VMOSFET, in turn, eliminates the need for amplification.

CHAPTER 10

Large-Scale Integration Devices

INTRODUCTION

The circuits of the previous chapters were designed with high current and high-power devices in mind. However, VMOSFETs are fabricated by large-scale integrated circuit technology. That is, a large number of individual short-channel vertical FETs are both manufactured and connected in parallel on the same "chip" of silicon.

The major advantage of the *vertical* MOSFET is its short channel, which is typically 1.5 μm. The layout reduces the total surface area of the silicon required for the manufacture of the device. This, in turn, suggests that these devices might be used to increase the number of active elements on an integrated circuit. In addition, the short channel produces faster switching times.

OBJECTIVES

In this chapter, the advantages of large-scale integrated circuits using vertical MOSFETs are presented.

VMOS RANDOM ACCESS MEMORIES

Among the advantages of VMOS technology are higher speed, lower power, and a reduction of chip area. Because of these characteristics, high-performance random access memories are now possible.

One is the S4016/2114 4K RAM, the pin diagram of which is shown in Fig. 10-1. It is arranged in a 1K (1024) by 4-bit configuration. Made by American Microsystems, Inc., the S4016/2114 features a 150-ns maximum access time while requiring only a +5-volt supply at 50 mA maximum. In addition, it has an output short-circuit current of 100 mA.

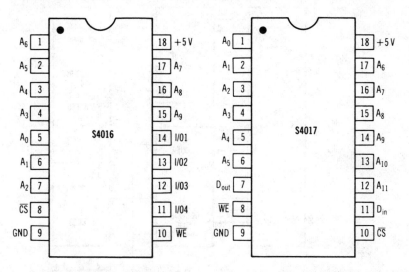

Fig. 10-1. Pin diagram, S4016
1K × 4 static RAM.

Fig. 10-2. Pin diagram, S4017
4K × 1 static RAM.

Similar to the S4016, the S4017 (Fig. 10-2) is also a 4K RAM, but arranged in a 4K (4096) by 1-bit configuration.

READ-ONLY MEMORY

The majority of the more complex microprocessor applications require between 2K and 8K of memory storage. To help meet this demand, the S4264 is a 64K ROM that is organized in an 8K (8192) by 8-bit configuration, because the 8 bit is the most popular, and 8K pages of words satisfy most needs (Fig. 10-3). It requires only a +5-volt supply and has TTL tri-state outputs, typical access times of 250 ns, and power dissipation of 500 mW.

For a typical 8-bit microprocessor system requiring 8K of program memory, such as the 6800, Fig. 10-4 shows a conventional arrangement using four 16K ROMs. Fig. 10-5 illustrates how the S4264 64K ROM can be used instead, which results in a considerable space savings and reduced component count.

SUMMARY

The VMOSFET is a technological breakthrough whose time has come. While no one has yet suggested that VMOSFETs can completely replace junction transistors, there are many advantages to designing new circuits utilizing VMOSFETs wherever possible. The

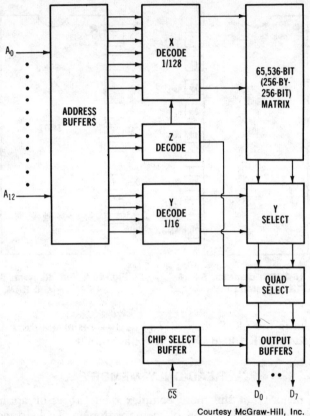

Courtesy McGraw-Hill, Inc.

Fig. 10-3. S4264 block diagram.

technology is no longer unproven; it is a reliable, widely sourced device, ready for increased demand and application. Failure to commit designs with this device can produce tangible and intangible disadvantages in a consumer industry where competition burns fiercely.

Rejection based on "retraining costs and delays" is unwarranted because of the simplicity of device models and their reliability in commercial products.

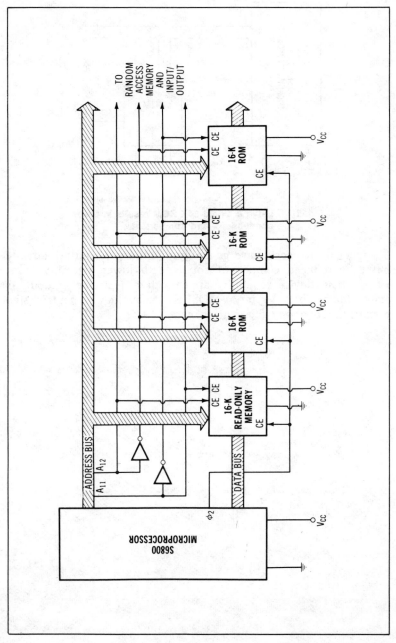

Courtesy McGraw-Hill, Inc.

Fig. 10-4. 6800 microprocessor with 16K ROMs.

101

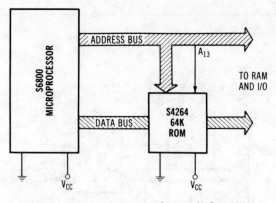

Courtesy McGraw-Hill, Inc.

Fig. 10-5. Using a single 64K ROM results in fewer connections and components.

Higher device costs have generally disappeared, and are already offset by reduced component counts with lower assembly costs. In addition, device reliability appears to be extremely good. As stated in the first chapter, VMOSFET technology is a giant step toward the ideal active circuit element.

CHAPTER 11

Performing the Experiments

OBJECTIVES

In this chapter you will learn:

- How to measure the threshold voltage and transconductance of a VMOSFET.
- The characteristics of the gate-protecting diode of a VMOSFET.
- The output characteristics of the drain of a VMOSFET.
- The ohmic region (nonsaturation) characteristics of a VMOS-FET.
- The characteristics of a VMOSFET common-source amplifier.
- The characteristics of a VMOSFET common-drain amplifier.
- The characteristics of two VMOSFETs operating in parallel.
- The characteristics of two VMOSFETs operating in series.
- The characteristics of VMOSFET-TTL interfacing.

RULES FOR SETTING UP EXPERIMENTS

In this chapter, you will have the opportunity to breadboard a number of circuits either using a variety of commercial breadboarding aids or constructing some of the necessary equipment. Before you set up any experiment, it is recommended that you do the following:

1. Plan your experiment beforehand. Know what types of results you are expected to observe.
2. Disconnect, or turn off, *all* power and external signal sources from the breadboard.

3. Clear the breadboard of all wires and components from previous experiments unless instructed otherwise.
4. Check the wired-up circuit against the schematic diagram to make sure that it is correct before starting.
5. Connect, or turn on, the power and external signal sources to the breadboard *last!*
6. When you are finished, make sure that you disconnect everything *before* you clear the breadboard of wires and components.

FORMAT FOR THE EXPERIMENTS

The instructions for each experiment are presented in the following format:

Purpose

The material presented under this heading states the purpose of performing the experiment. It is well for you to have this intended purpose in mind as you conduct the experiment.

Pin Configuration

The pin configuration of the solid-state device is presented under this heading.

Schematic Diagram of Circuit

The schematic diagram of the completed circuit that you will wire up in the experiment is presented under this heading. You should analyze this diagram in an effort to obtain an understanding of the circuit *before* you proceed further.

Design Basics

A summary of the design equations and/or characteristics that apply to the design and operation of the circuit is given under this heading.

Steps

A series of sequential steps that describe the detailed instructions for performing portions of the experiments is presented here. Any numerical calculations are performed easily on many of the simple pocket-type calculators.

HOW MANY EXPERIMENTS DO I PERFORM?

In this chapter, there are a number of experiments. While it is not necessary to perform all the experiments, you are encouraged to do as many as possible to gain a good feeling for the operation of a number of the VMOS parameters and circuits.

EQUIPMENT

The following pieces of equipment will be required to perform the experiments:

Power Supplies

Two well-filtered dc supplies are required. One is a variable 24-volt supply with a capacity of supplying 2 amperes, while the other must supply 100 volts at 1 ampere.

Meters

For voltage measurements, a digital voltmeter is preferred for precise measurements.

For current measurements, two meters, each capable of measuring 10 mA, are required.

Oscilloscope

Just about any general-purpose type will do, although a dual-trace type is preferred.

Signal Generator

A variable frequency (up to 100 kHz) variable output signal generator capable of generating low-distortion sine waves is required.

Precision Pulse Generator

In order to accurately demonstrate the measurement of several parameters and circuits, a variable pulse width, variable output voltage generator is required. A simple circuit, shown in Fig. 11-1, uses a 555 timer and provides a continuously variable pulse amplitude from 0 to +10 volts with nearly a 1% duty cycle at a 100-Hz repetition rate.

The precision pulse generator must be calibrated prior to use, and the power supply to the generator must be well regulated.

R_4 of the precision pulse generator must be a 10-turn precision dial potentiometer.

To calibrate the pulse generator after construction is complete, ground the gate of Q_1 and adjust R_3 to obtain a drain voltage of +10 volts to ground on Q_1. Disconnect the ground from the gate of Q_1.

The output amplitude of the pulse generator should now be equal to the dial setting of the 10-turn precision potentiometer.

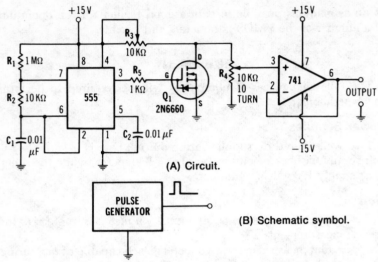

(A) Circuit.

(B) Schematic symbol.

Fig. 11-1. Precision pulse generator using a 555 timer.

Examine the output waveform using an oscilloscope. You should observe an approximately rectangular pulse of about 80 μs to 100 μs. duration. The pulse repetition rate should be about 100 Hz.

You should be able to determine the pulse amplitude using the pulse catcher and a DVM. However, the pulse catcher cannot correctly detect pulses greater than about 7 volts without increasing its supply voltage.

Precision Pulse Catcher

For testing devices capable of handling large currents and voltages, a significant amount of power must be dissipated if all areas of operation are to be examined. A technique that avoids these disadvantages is to perform certain tests in a pulsed mode of operation.

The precision pulse catcher, a schematic of which is shown in Fig. 11-2, is a modified peak detector using an operational amplifier. It is used in conjunction with the pulse generator of Fig. 11-1.

COMPONENTS

A list of all the various components needed to perform all the experiments given in this chapter is shown in Table 11-1.

AN INTRODUCTION TO THE EXPERIMENTS

The following experiments are designed to illustrate the measurement of several VMOS device parameters and amplifier circuits.

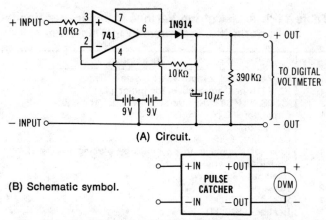

(A) Circuit.

(B) Schematic symbol.

Fig. 11-2. Precision pulse catcher with a digital voltmeter.

The experiments that you will perform can be summarized as follows:

Experiment No.	Purpose
1	Measure the threshold voltage and transconductance of a VMOSFET.
2	Determine the characteristics of the gate protective diode of a VMOSFET.
3	Determine the incremental drain resistance of a VMOSFET.
4	Demonstrate the characteristics of the non-saturation region of a VMOSFET.
5	Demonstrate the operation of a VMOSFET common-source amplifier.
6	Demonstrate the operation of a VMOSFET common-drain amplifier.
7	Determine the current handling capacity of two VMOSFETs connected in parallel.
8	Demonstrate the characteristics of two VMOSFETs connected in series.
9	Determine the gate voltage enhancement requirements of a VMOSFET.

EXPERIMENT NO. 1

Purpose

The purpose of this experiment is to determine the threshold voltage V_T and the transconductance g_{fs} of a type VN46AF VMOSFET.

Table 11-1. A List of the Various Components Needed to Perform the Experiments in This Chapter

Quantity	Description
1	IC, 7400 quad 2-input NAND gate (TTL)
1	IC, 7416 hex open collector buffer (TTL)
2	VMOS, VN46AF *n*-channel power FET (or equivalent)
1	Capacitor, 39 pF
1	Capacitor, 50 pF
1	Capacitor, 1 μF
1	Resistor, 1 ohm (can be made by placing twelve 12-ohm resistors in parallel)
1	Resistor, 24 Ω, 6 W
2	Resistor, 1 kΩ, ½ W
1	Resistor, 720 kΩ, ½ W
1	Resistor, 820 kΩ, ½ W
1	Resistor, 1 MΩ, ½ W
1	Resistor, 1.5 MΩ, ½ W
1	Resistor, 39 MΩ, ½ W

Pin Configuration of VMOSFET (Fig. 11-3)

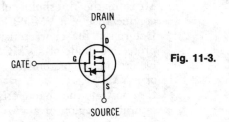

Fig. 11-3.

Schematic Diagram of Circuit (Fig. 11-4)

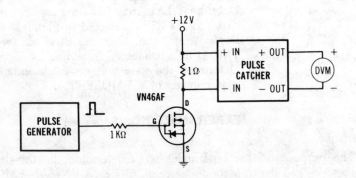

Fig. 11-4.

Design Basics

Transconductance:

$$g_{fs} = \frac{\Delta I_D}{\Delta V_{GS}}$$

where,

g_{fs} is transconductance,
I_D is drain current,
V_{GS} is gate-source voltage.

Step 1

With the power off, wire the circuit as shown in the schematic, making sure that the leads of the VMOSFET are connected correctly and kept as short as possible. Also, connect the 1-kΩ resistor as close as possible to the gate lead of the VMOSFET.

Step 2

Apply power to the breadboard and connect the output of the precision pulse generator to the input of the circuit. Initially, adjust the output of the generator to zero volts.

Step 3

Now vary the input voltage from the pulse generator, measuring the gate-source voltage V_{GS} and the drain current I_D and recording your results in the following table. The drain current is determined by measuring the output voltage of the pulse catcher, which is the voltage across the 1-ohm resistor, and after application of Ohm's law:

$$I_D = (V_{\text{pulse catcher}})/(1\,\Omega)$$

V_{GS} (measured)	I_D (calculated)
1 V	
2 V	
3 V	
4 V	
5 V	
6 V	
7 V	
8 V	
9 V	
10 V	

Step 4

On the graph in Fig. 11-5, plot the calculated drain current (vertical axis) versus the measured gate-source voltage and draw a smooth curve through all the points.

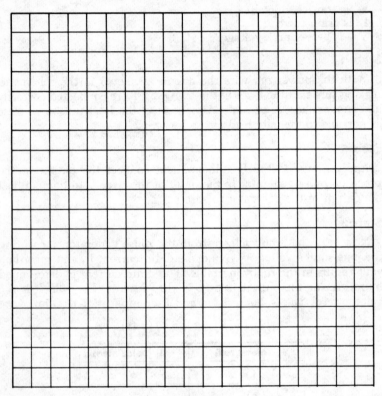

Fig. 11-5.

Step 5

Now draw a straight line through the points on the curve at which V_{GS} equals 5 volts and 10 volts; extend this line until it intersects with the V_{GS} axis. This point of intersection is termed the *threshold voltage*, V_T, and represents the minimum gate-source voltage necessary to ensure that a VMOSFET is in the operating, or conducting, region.

Step 6

From the transconductance formula given in the "Design Basics" section, determine the transconductance of the VMOSFET used for this experiment for each of the points you determined in Step 3 and record your results in the following table. For the first point, as an example, the transconductance is the ratio of the change in current to the gate-source voltage change from 0 to 1 volt. The next point is the change in current as the voltage increases from 1 volt to 2 volts, and so on.

ΔV_{GS}	ΔI_D	g_{fs}
0–1 V		
1–2 V		
2–3 V		
3–4 V		
4–5 V		
5–6 V		
6–7 V		
7–8 V		
8–9 V		
9–10 V		

Step 7

Now plot the transconductance, in mhos, versus the gate-source voltage on the graph in Fig. 11-6. Our results looked very similar to the VMOS curve in Fig. 1-3. Threshold voltage (V_T) was 2.6 volts, and g_{fs} was 0.294 mhos, using a 2N6657 transistor.

Your plot should show that the transconductance increases until the gate-source voltage is approximately 4 volts to 5 volts, after which the transconductance remains essentially constant for further increases in V_{GS}. This is due to the construction of the VMOS device which produces the greatest drop in drain-source voltage across the epitaxial layer of the transistor. A drain-source voltage change cannot affect the channel velocity of the charge carriers, so that the "square law" current characteristic of standard planar field-effect transistors is not apparent. VMOSFETs are linear devices for most values of gate-source voltages.

EXPERIMENT NO. 2

Purpose

The purpose of this experiment is to determine the characteristics of the gate protective diode of a type VN46AF VMOSFET.

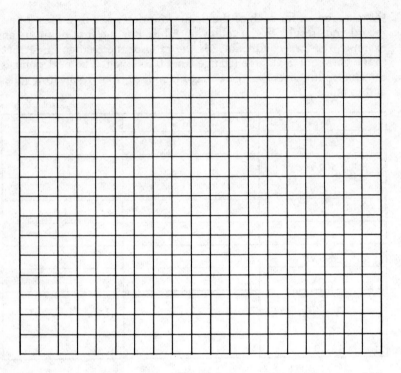

Fig. 11-6.

Pin Configuration of VMOSFET (Fig. 11-7)

Schematic Diagram of Circuit (Fig. 11-8)

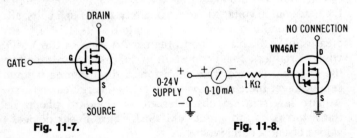

Fig. 11-7.   Fig. 11-8.

Step 1

Wire the circuit as shown in the schematic diagram. As in the previous experiment, try to keep the leads of the VMOSFET as short as possible. Initially, leave the power supply disconnected.

Step 2

Turn on the variable power supply and set its output voltage at zero volts. Now connect the supply to the remainder of the circuit.

Step 3

Now slowly increase the power supply voltage until the ammeter begins to indicate current flow (about 1 mA). Measure the gate-source voltage and record it.

$$V_{GS}(1) = \text{\_\_\_\_\_} \text{ volts}$$

Step 4

Increase the supply voltage until the gate current equals 5 mA, and measure the resulting gate-source voltage.

$$V_{GS}(2) = \text{\_\_\_\_\_} \text{ volts}$$

Step 5

Decrease the power supply voltage to zero and reverse its connections so that the positive lead is now grounded. In addition, reverse the leads of the milliammeter.

Step 6

Now increase the power supply voltage until the current reaches 5 mA. Measure the resultant gate-source voltage.

$$V_{GS}(3) = \text{\_\_\_\_\_} \text{ volts}$$

Can you explain the difference between this value and that measured in Step 4?

This is the forward bias voltage of the gate protection "diode." Step 4 measured the sustaining reverse breakdown voltage of the gate protecting diode.

Step 7

With the gate voltage supply kept unchanged, disconnect this voltage from the circuit and place a second milliammeter in series with the drain of the VMOSFET in addition to a second power supply, as shown in Fig. 11-9.

Step 8

Adjust the drain supply voltage at +12 volts and connect the gate supply voltage to the circuit. Now adjust the gate supply until the gate current equals 5 mA and record the corresponding drain current.

$$I_D(I_G \text{ @ } 5\,\text{mA}) = \text{\_\_\_\_\_}$$

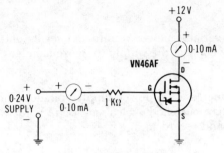

Fig. 11-9.

We measured 1.5 mA. What has happened? This indicated that the gate protecting transistor has a fairly low forward current transfer ratio.

The protected gate VMOSFET is constructed in such a manner that it acts as an npn transistor (assuming that an *n*-channel device is used) when the gate is negative with respect to its source. When this occurs, a current limiting resistor should be placed in series with the gate to prevent excessively large currents from destroying the device.

Step 9

If available, repeat this experiment using an unprotected gate VMOSFET, such as a VN64GA or VN66AJ. Do you notice any difference in your measurements?

EXPERIMENT NO. 3

Purpose

The purpose of this experiment is to determine the incremental drain resistance of a VN46AF VMOSFET.

Pin Configuration of VMOSFET (Fig. 11-10)

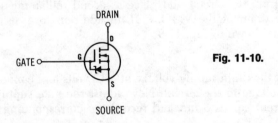

Fig. 11-10.

Schematic Diagram of Circuit (Fig. 11-11)

114

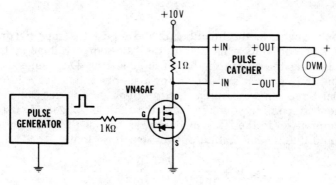

Fig. 11-11.

Design Basics

Incremental drain (output) resistance:

$$r_o = \frac{\Delta V_{\text{supply}}}{\Delta I_D}$$

Step 1

Wire the circuit as shown in the schematic diagram. As in Experiment No. 1, the drain current will be measured by measuring the output voltage of the precision pulse catcher and applying Ohm's law.

V_{GS}	I_D (+10 V)	I_D (+30 V)	r_o
1 V			
2 V			
3 V			
4 V			
5 V			
6 V			
7 V			
8 V			
9 V			
10 V			

Step 2

Apply power to the breadboard and adjust the supply (drain) voltage at +10 volts while setting the gate-source voltage at 1 volt by adjusting the output of the pulse generator. Determine the resultant drain current and repeat for a supply voltage of +30 volts. Repeat these same measurements for 1-volt gate-source voltage increments, recording your results in the table on page 115. Use the equation given in the "Design Basics" section of this experiment to compute the incremental output resistance, so that

$$r_o = \frac{30\,V - 10\,V}{I_D(30\,V) - I_D(10\,V)}$$

The output resistance of your VMOSFET should be greater than 10 kΩ, which is quite high and is very similar to the characteristics of a pentode vacuum tube. The relatively constant current ouput with varying supply voltage suggests that the VMOSFET is ideal as a constant current source, which is described in Chapter 9. Keep this circuit wired for the next experiment.

EXPERIMENT NO. 4

Purpose

The purpose of this experiment is to demonstrate the characteristics of the nonsaturation region of a VN46AF VMOSFET.

Pin Configuration of VMOSFET (Fig. 11-12)

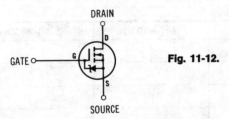

Fig. 11-12.

Design Basics

Drain resistance:

$$r_{DS} = \frac{\Delta V_{DS}}{\Delta I_D}$$

Step 1

Wire the circuit as shown in the schematic diagram, unless you have just performed the previous experiment, and set the drain

Schematic Diagram of Circuit (Fig. 11-13)

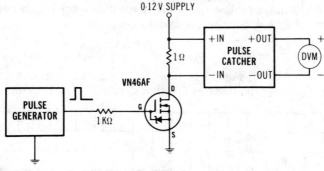

Fig. 11-13.

supply voltage V_{DS} to zero without connecting this power to the breadboard.

Step 2

Now apply power to the breadboard. Using an oscilloscope, adjust the output peak voltage of the precision pulse generator to 1 volt, which is also the gate-source voltage, V_{GS}. Using the pulse

	Measured Drain Current I_D @ Gate-Source Voltage V_{GS} =									
V_{DS}	1 V	2 V	3 V	4 V	5 V	6 V	7 V	8 V	9 V	10 V
1 V										
2 V										
3 V										
4 V										
5 V										
6 V										
7 V										
8 V										
9 V										
10 V										
11 V										
12 V										

catcher to measure the drain current, I_D, measure the drain current while varying the drain supply voltage from 0 to +12 volts, recording your results in the table on page 117. Repeat this step for gate-source voltages of 2, 3, 4, 5, 6, 7, 8, 9, and 10 volts.

Step 3

On the graph in Fig. 11-14 plot the results that you obtained in Step 2. Your plot should resemble that shown in Figs. 1-4A and 1-4B.

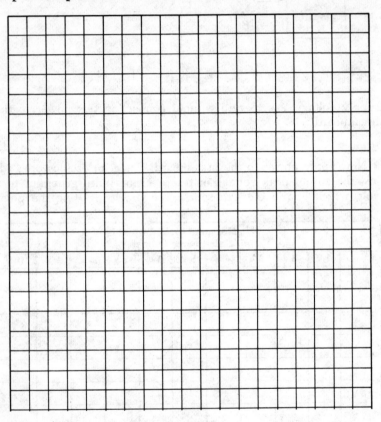

Fig. 11-14.

Step 4

Using the equation given in the "Design Basics" section of this experiment, calculate the drain resistance of your VMOSFET for the linearly increasing region of the graph, using a gate-source voltage of 5 volts. For example, our results showed

$$r_{DS}(V_{GS} = 5\text{ V}) = \frac{\Delta V_{DS}}{\Delta I_D}$$
$$= \frac{(1.0 - 0.5)\text{ volts}}{(0.5 - 0.25)\text{ amps}}$$
$$= 2\text{ ohms}$$

EXPERIMENT NO. 5

Purpose

The purpose of this experiment is to demonstrate the operation and characteristics of a VMOSFET common-source amplifier.

Pin Configuration of VMOSFET (Fig. 11-15)

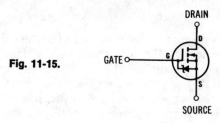

Fig. 11-15.

Schematic Diagram of Circuit (Fig. 11-16)

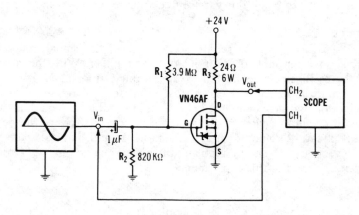

Fig. 11-16.

Design Basics

Voltage gain:

$$A_V = \frac{V_o}{V_i}$$

Amplifier Distortion:

$$\%D = \frac{I_D max + I_D min - 2I_Q}{2(I_D max - I_D min)} \times 100$$

where,

$I_D min$ is minimum output (drain) current,
$I_D max$ is peak output current,
I_Q is output current with no input voltage; i.e., quiescent current.

Step 1

Wire the circuit as shown in the schematic diagram. Because of the potentially large amount of power that will be dissipated, the transistor should be mounted on a finned heat sink approximately 5 inches square.

Step 2

Leave the signal generator disconnected from the circuit and apply power to the circuit. With the oscilloscope connected to the drain of the transistor, the quiescent output voltage should be 12 volts ± 2 volts dc. If it is not, disconnect the power from the circuit and first check your wiring against the schematic diagram. If everything is correct, repeat Experiment No. 1 with this transistor and determine the gate-source voltage, V_{GS}, required to produce 0.5 ampere drain current. Increase or decrease the value of resistor R_2 as necessary to reach this gate-source voltage.

Step 3

When this circuit is set properly, connect the signal generator (sine wave output) to the input of the amplifier. With the probes of the oscilloscope connected to the output of the signal generator (V_i) and the drain of the transistor (V_o), set the frequency of the generator at approximately 1 kHz. Using the table on page 121 and recording your results, adjust the input voltage to produce the peak-to-peak output voltages listed in the table. In addition, calculate the voltage gain of the amplifier using the equation given in the "Design Basics" section of this experiment.

Step 4

Now disconnect the signal generator from the input of the amplifier. Measure the quiescent drain current and record your result.

$$I_Q = \underline{\hspace{2cm}}$$

Step 5

Reconnect the signal generator to the input of the amplifier and adjust the output of the generator at 1 volt peak-to-peak at a fre-

V_1	V_o	A_V
	1 V	
	2 V	
	3 V	
	4 V	
	5 V	
	6 V	
	7 V	
	8 V	
	9 V	
	10 V	

quency of 1 kHz. Using the pulse catcher, measure the drain, or output current, which will be $I_D$max, and record your result.

$$I_D\text{max} = \underline{\hspace{1cm}}$$

Step 6

Disconnect the signal generator and then the power supply from the circuit. Then insert a 1-μF capacitor in series with the *positive* polarity input of the pulse catcher, as shown in Fig. 11-17.

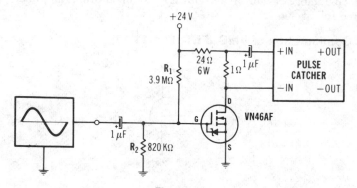

Fig. 11-17.

Step 7

Reconnect the power and the signal generator to the circuit and measure the output current, using the pulse catcher, with the same settings outlined in Step 5, and record the peak-to-peak output current.

$$I_Dp\text{-}p = \underline{\hspace{1cm}}$$

Step 8

To determine $I_D$min, subtract the value determined in Step 7 from the value determined in Step 5.

$$I_D\text{min} = I_D\text{max} - I_Dp\text{-}p = \underline{\hspace{1cm}}$$

Step 9

From the equation given in the "Design Basics" section of this experiment, calculate the percent distortion of this common-source amplifier at 1 kHz.

$$\%\ \text{distortion} = \underline{\hspace{1cm}}$$

Our results showed a 5.5% distortion.

Step 10 (optional)

Repeat Steps 5 through 9 with peak-to-peak output voltages from 1 volt to 10 volts.

EXPERIMENT NO. 6

Purpose

The purpose of this experiment is to demonstrate the operation and characteristics of a VMOSFET common-drain amplifier, or source follower.

Pin Configuration of VMOSFET (Fig. 11-18)

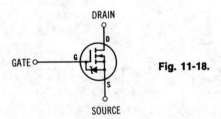

Fig. 11-18.

Step 1

Wire the circuit as shown in the schematic diagram. As in the previous experiment, the VMOSFET should be mounted on a suitable heat sink.

Schematic Diagram of Circuit (Fig. 11-19)

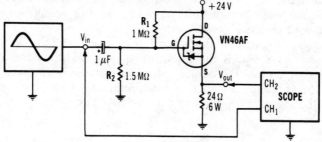

Fig. 11-19.

Step 2

First leave the signal generator disconnected, and apply power to the circuit. With the oscilloscope connected to the source of the transistor, the quiescent output voltage should be 12 volts ± 2 volts dc. If it is not, disconnect the power from the breadboard and first check your wiring against the schematic diagram. If everything is correct, increase or decrease the value of resistor R_2 as necessary to set the quiescent source voltage output at 12 volts ± 2 volts dc.

Step 3

When this circuit is set properly, connect the signal generator (sine wave output) to the input of the circuit. With the probes of the oscilloscope connected to the output of the signal generator (V_i) and the source of the transistor (V_o), set the frequency of the

V_I	V_o	A_V
1 V		
2 V		
3 V		
4 V		
5 V		
6 V		
7 V		
8 V		
9 V		
10 V		

generator at approximately 1 kHz. Now set the peak-to-peak input voltage according to the table on page 123, measure the corresponding peak-to-peak output voltage, determine the voltage gain, and record your results.

You should find that the voltage gain is one, so that the output voltage always equals the input voltage. In addition, there is no phase shift, so that the output voltage is essentially an exact representation of the input voltage. Consequently, the source voltage "follows" the input, hence the name *source follower*.

Step 4

Disconnect both the power and the signal generator from the circuit. As shown in Fig. 11-20, place a 1-ohm resistor in series with the source resistor.

Step 5

Connect only the power to the circuit and, using the pulse catcher, measure the quiescent source current and record your result.

$$I_Q = \underline{\qquad}$$

Step 6

Reconnect the signal generator to the input of the circuit and adjust the output of the generator at 1 volt peak-to-peak at a frequency of 1 kHz. Using the pulse catcher, measure the maximum source current and record your result.

$$I_S max = \underline{\qquad}$$

Step 7

Disconnect the signal generator and then insert a 1-μF capacitor in series with the *positive* polarity input of the pulse catcher, as was shown in Fig. 11-17 for the previous experiment.

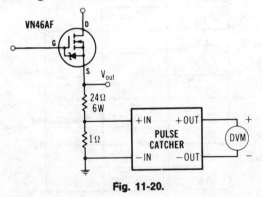

Fig. 11-20.

Step 8

Reconnect the signal generator to the circuit and measure the peak-to-peak source current, using the same settings in Step 6, and record your result.

$$I_Sp\text{-}p = \underline{\hspace{1.5cm}}$$

Step 9

To determine the minimum source current, subtract the value determined in Step 8 from the value determined in Step 6.

$$I_Smin = I_Smax - I_Sp\text{-}p = \underline{\hspace{1.5cm}}$$

Step 10

Now calculate the percent distortion of your source follower using the following formula:

$$\% \text{ distortion} = \frac{I_Smax + I_Smin - 2I_Q}{2(I_Smax - I_Smin)} \times 100 = \underline{\hspace{1.5cm}}$$

Distortion was too low to measure using this approach in our experiment.

You should note that the distortion of this source follower is quite low, and less than the common-source amplifier of the previous experiment.

Step 11 (optional)

Repeat Steps 5 through 10 with peak-to-peak input voltages ranging from 1 volt to 10 volts.

EXPERIMENT NO. 7

Purpose

The purpose of this experiment is to determine the current handling capacity of two VMOSFETs connected in parallel.

Pin Configuration of VMOSFET (Fig. 11-21)

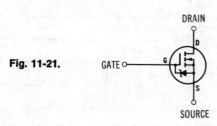

Fig. 11-21.
GATE
DRAIN
G
D
S
SOURCE

Schematic Diagram of Circuit (Fig. 11-22)

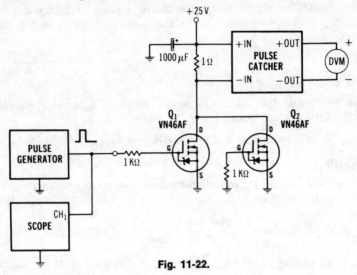

Fig. 11-22.

Step 1

Wire the circuit as shown in the schematic diagram. Initially ground the 1-kΩ resistor that is connected to the gate of VMOSFET Q_2 and connect the precision pulse generator to the input circuit of Q_1.

V_{GS}	VMOS Q_1 I_{D1}	VMOS Q_2 I_{D2}	$I_{D1} + I_{D2}$	Measured I_D total
1 V				
2 V				
3 V				
4 V				
5 V				
6 V				
7 V				
8 V				
9 V				
10 V				

Step 2

Apply power and the signal from the pulse generator to the circuit. Using an oscilloscope, adjust the peak output voltage of the precision pulse generator at 1 volt, which is also the gate-source voltage, V_{GS}.

Using the pulse catcher to measure the drain current, I_{D1}, record your results in the table on page 126, and then repeat the process for gate-source voltages up to 10 volts.

Step 3

Repeat Step 2 by connecting the pulse generator to the input of Q_2 and then grounding the input circuit of Q_1. Measure the drain current I_{D2} and record your results in Column 3 of the table. Then add the values for I_{D1} and I_{D2} for each gate-source voltage, and record your results in Column 4 of the table.

Step 4

Now connect both inputs together so that they are both connected to the precision pulse generator.

As in the previous steps, measure the total drain current, but now of both devices in parallel for all the values of V_{GS} listed in the table.

Record your values and compare them with the corresponding values in Column 4.

The values in Column 4 should, ideally, equal those corresponding values of Column 5.

Step 5

If the transconductance of each VMOSFET is 0.25 mho, what would be the effective transconductance of two VMOSFETs operating in parallel, as in this experiment?

The effective transconductance of two devices operating in parallel is twice the value of a single device, so that

$$g_{fs}(\text{effective}) = 2g_{fs}$$
$$= (2)(0.25)$$
$$= 0.5 \text{ mho}$$

If the "on" channel, or drain-source resistance of each VMOSFET is 2 ohms, then what is the effective drain-source resistance of two devices operating in parallel?

The effective drain-source resistance r_{DS} of two like devices operating in parallel is one-half the value of a single device, or

$$r_{DS}(\text{effective}) = \frac{r_{DS}}{2}$$

$$= \frac{2}{2}$$

$$= 1 \text{ ohm}$$

EXPERIMENT NO. 8

Purpose

The purpose of this experiment is to demonstrate the characteristics of VMOSFETs operating in series.

Pin Configuration of VMOSFET (Fig. 11-23)

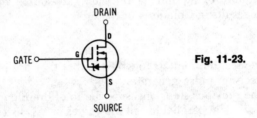

Fig. 11-23.

Schematic Diagram of Circuit (Fig. 11-24)

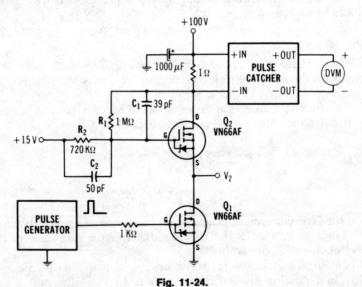

Fig. 11-24.

Design Basics

Dynamic balance:

$$\frac{C_1}{C_2} = \frac{R_2}{R_1}$$

Gate bias voltage:

$$V_G(Q_2) = 15 + \frac{(V_G(Q_2) - 15)R_2}{R_1 + R_2}$$

Step 1

Wire the circuit as shown in the schematic diagram. You are strongly cautioned that the 100-volt supply can be deadly. *Make all connections with the power off; turn off the power supply when moving the voltmeter leads; and do not work on a conducting surface or a grounded floor.*

Step 2

With the pulse generator output voltage at zero, apply power to the circuit. Measure voltage V_2 (i.e., the drain voltage of Q_1) and record it.

$$V_2 = \underline{\hspace{1cm}} \text{ volts}$$

If everything is working correctly, this voltage should be approximately +50 volts, which is one-half the supply voltage. This is so that the drain-source voltage drops are equal with no input signal.

Step 3

With the gate-source voltages listed in the table on page 130, which are set by the output pulse heights of the pulse generator, measure the resultant drain currents. If you are using the same VMOSFETs that were used in Experiment No. 1, copy these results for the corresponding gate-source voltages in the last column.

The two drain currents should be equal.

EXPERIMENT NO. 9

Purpose

The purpose of this experiment is to demonstrate the interfacing of a VMOSFET with TTL devices, such as a 7400 NAND gate and a 7416 inverter/buffer.

		Experiment No. 1
V_{GS}	I_D	I_D
1 V		
2 V		
3 V		
4 V		
5 V		
6 V		
7 V		
8 V		
9 V		
10 V		

Pin Configuration of Integrated Circuit Chips and VMOSFET (Fig. 11-25)

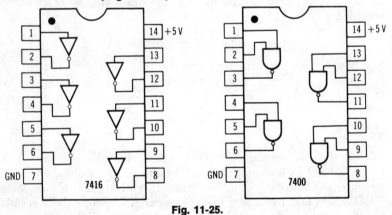

Fig. 11-25.

Step 1

Wire the circuit as shown in the schematic diagram. Don't forget the supply connections to the 7400 NAND gate integrated circuit; connect pin 14 to +5 volts, and pin 7 to ground.

Step 2

Apply power to the circuit and initially leave the 1-kΩ pull-up resistor, which is the one shown connected from the output of the second NAND gate to +5 volts, disconnected from the circuit. Using the pulse catcher, measure the drain current of the VMOSFET and record your result.

$$I_D = \underline{\qquad}$$

Schematic Diagram of Circuit (Fig. 11-26)

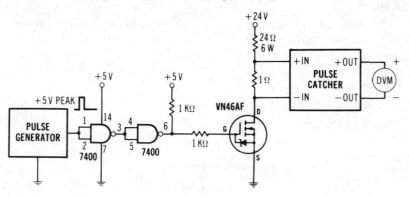

Fig. 11-26.

Step 3

Connect the 1-kΩ pull-up resistor as shown in the schematic diagram. Now measure the drain current of the VMOSFET. Your value should be different from that obtained in Step 2. Why?

Since the pull-up resistor now provides a logic 1 level of +5 volts, instead of about 4.3 volts without it, the drain current is, therefore, greater with the pull-up resistor.

Step 4

Disconnect the power from the breadboard and wire the input circuit as shown in Fig. 11-27, using a 7416 open-collector inverter/ buffer. Don't forget the power supply connections for the 7416; connect pin 14 to +5 volts, and pin 7 to ground.

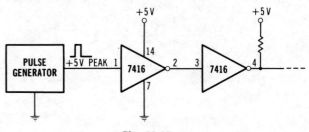

Fig. 11-27.

Step 5

Apply power to the circuit, measure the VMOSFET drain current, and record your result.

$$I_D = \underline{\qquad}$$

Step 6

Now reduce the +15-volt supply connected to the pull-up resistor in 1-volt increments to +5 volts; measure the resulting drain current and record your results in the following table:

V	I_D
15 V	
14 V	
13 V	
12 V	
11 V	
10 V	
9 V	
8 V	
7 V	
6 V	
5 V	

Glossary

base—The controlling region of a junction transistor. The base separates the emitter and collector.

BJT—Base junction transistor in which the output current is controlled by the input current.

bonding—Attachment. Electrons are said to be bound to other atoms or have shared bonding with two atoms.

channel—A current-carrying layer that forms between the source and drain of a field-effect transistor.

collector—The region of a junction transistor where the output current is collected.

crystal—A solid form of matter in which the atoms are tightly bonded together by the sharing of electrons with neighboring atoms. Silicon crystals are a cubic structure in which each atom shares four electrons with four neighboring atoms.

depletion FET—A field-effect transistor that will conduct current with no voltage applied between the gate and source terminals. The current through the device may be increased or decreased by varying the gate-source voltage.

diffusion—The random movement of particles from an area of high concentration to an area of lower concentration.

diffusion capacitance—A change in the charge stored within a region as the current through that region increases. To increase diffusion current flow requires an increase in the difference in concentration of charge carriers from a region of high to low concentration.

diode—A two-terminal device consisting of an *n*, or electron donor region, and a *p*, or electron acceptor region, which are both physically bonded together.

doping—The addition of impurities with either five (for *n*-type) or three (for *p*-type) outer electrons.

drift—The movement of particles under the influence of an external force.

electric field—A region of push or pull on electrically charged particles.

electron—A fundamental particle of an atom. The electron has a negative charge. Combinations of outer electrons of atoms give rise to crystals and molecules.

electron acceptor—A semiconductor impurity with three outer electrons.

electron donor—A semiconductor impurity with five outer electrons.

emitter—The region of a junction transistor that is the source of the current through the transistor.

enhancement FET—A field-effect transistor that will not conduct current unless a voltage is applied between the gate and source terminals.

feedback—The return of a fraction of the output signal of a circuit to its input.

field-effect transistor (FET)—A transistor whose current flow is controlled by the electric field on the control lead, or *gate*. The terminals of a FET are source, which is the source of current through the FET; gate, which is the current controlling terminal; and drain, which is the region where current carriers are collected.

forward bias—The condition when negative voltage is applied to the *n*-region and positive voltage is applied to the *p*-region.

gate-drain capacitance—The capacitance that appears between the gate and drain terminals in a FET, and is abbreviated C_{gd}.

gate-source capacitance—The capacitance that appears between the gate and source terminals of a FET, and is abbreviated C_{gs}.

I_D *(drain current)*—The current entering the drain terminal of a FET.

I_G *(gate current)*—The current entering the gate terminal of a FET.

I_S *(source current)*—The current entering the source terminal of a FET.

injection—Supplying charge carriers into a semiconductor region.

minority carriers—When electron and hole charge carriers are transferring electrical energy within the same region, the carriers with the lowest concentration are said to be minority carriers. Electrons are minority carriers in a *p*-type semiconductor. Holes are minority carriers in an *n*-type semiconductor.

mobility—The ease of movement of electrons or holes in a crystal.

MOSFET—Metal oxide semiconductor field-effect transistor. A field-effect transistor device whose gate lead is insulated from the body of the transistor by a layer of metal oxide.

n-channel FET—A field-effect transistor designed to use electrons as the current carriers through the transistor.

n-type semiconductor—A semiconductor, which may be either germanium or silicon, that has a small number of atoms with five outer electrons, as compared with the four outer electrons of the semiconductor. The extra electron is not included in the crystal bonding, and is free to move.

negative feedback—The return of a portion of the output signal of a circuit to its input, fed back 180° out of phase with the input signal. Consequently, this type of feedback reduces circuit gain.

npn transistor—A junction transistor that uses electrons as the major charge carriers. The base of the transistor is made from a *p*-type material, and separates the emitter and collector, which are both made from *n*-type material.

p-channel FET—A field-effect transistor designed to use holes as the current carriers through the transistor.

p-type semiconductor—A semiconductor that has a small number of atoms with three outer electrons, as compared with the four outer electrons of the semiconductor. The missing electron is readily replaced with an electron from a neighboring atom, resulting in a movement of the missing electron hole.

planar MOSFET—A field-effect transistor built on the surface of a piece of semiconductor material. All electrical connections are on the same surface of the material.

pnp transistor—A junction transistor that uses holes as the major charge carriers. The base of the transistor is made from an *n*-type material, and separates the emitter and collector, which are both made from *p*-type material.

positive feedback—The return of a portion of the output signal of a circuit to its input, in phase with the input signal. Although this results in increased gain, it also can cause the circuit to oscillate.

recombination—The result of a free electron filling a hole. The hole and free electron charge carriers disappear by joining the crystal structure.

reverse bias—Negative voltage on the *n* region and positive voltage on the *p* region.

saturation—For junction transistors, saturation occurs when there are many more current carriers injected into the base region than can be collected by the collector of the transistor. The voltage between the emitter and collector is typically 0.2 volt when in saturation. For field-effect transistors, the carriers in the conducting region (channel) are moving at maximum velocity. As the drain voltage is increased in saturation, very little increase in current is observed.

space charge capacitance—The capacitance across a diode junction that results from the *n* region and *p* region (conducting areas), separated by a depletion region (insulator).

transconductance—*Trans*fer *conductance*. A measure of the change in output current of a device produced by a change in input voltage, and abbreviated either g_m or g_{fs}.

V_{DS} (*drain-source voltage*)—The voltage on the drain terminal measured with respect to the source terminal.

V_{GS} (*gate-source voltage*)—The controlling voltage on a FET.

VMOSFET—Vertical metal oxide semiconductor field-effect transistor. A field-effect transistor that has the source and drain arranged in a vertical sandwich.

Data Sheets

This appendix contains the following data sheets, which are reproduced with permission of their respective manufacturers:

Hitachi America, Ltd. 707 West Algonquin Road Arlington Heights, IL 60005	2SJ48/49/50 p-channel power MOSFETs 2SK133/134/135 n-channel power MOSFETs
International Rectifier Semiconductor Division 233 Kansas Street El Segundo, CA 90245	IRF 100/101 n-channel MOSFETs (high current) IRF 300/301/305/306 n-channel MOSFETs (high voltage)
ITT Semiconductors 500 Broadway Lawrence, MA 01841	BD 512 p-channel power MOSFET BD 522 n-channel power MOSFET BS 170 n-channel VMOS and BS 250 p-channel VMOS
SEMTECH Corporation 652 Mitchell Road Newbery Park, CA 91320	2N6656/57/58 n-channel VMOS power FETs 2N6659/60/61 n-channel VMOS power FETs

2SJ48/49/50
P-Channel Power MOSFETs

■ **FEATURES:**
- **Superior High Frequency Characteristics**
- **High Speed Switching Characteristics**
- **Superior Durability**
- **Good Complementary Characteristics with 2SK133/134/135**

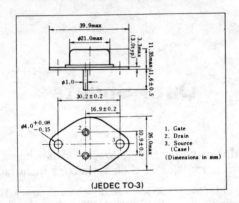

1. Gate
2. Drain
3. Source (Case)

(Dimensions in mm)

(JEDEC TO-3)

■ **ABSOLUTE MAXIMUM RATINGS** ($T_a = 25°C$)

Item	Symbol	Value	Unit
Drain to Source Voltage 2SJ48 2SJ49 2SJ50	V_{DSX}	−120 −140 −160	V
Gate to Source Voltage	V_{GSS}	±14	V
Drain Current	I_D	−7	A
Channel Dissipation	$P_{ch}*$	100	W
Storage Temperature	T_{stg}	−55~+150	°C

* Value at $T_C = 25°C$

MAXIMUM CHANNEL DISSIPATION CURVE

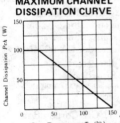

■ **ELECTRICAL CHARACTERISTICS** ($T_a = 25°C$)

Item	Symbol	Test Condition	min	typ	max	Unit
Drain to Source Breakdown Voltage 2SJ48 2SJ49 2SJ50	$V_{(BR)DSX}$	$I_D = -10mA$, $V_{GS} = 10V$	−120 −140 −160	− − −	− − −	V
Gate to Source Breakdown Voltage	$V_{(BR)GSS}$	$I_G = ±100μA$, $V_{DS} = 0$	±14	−	−	V
Gate to Source Cut-off Voltage	$V_{GS(off)}$	$I_D = -100mA$, $V_{DS} = -10V$	0	−0.8	−1.5	V
Drain to Source Saturation Voltage	$V_{DS(sat)}$	$I_D = -7A$, $V_{GD} = 0$	−	−	−12	V
Forward Transfer Admittance	$|yfs|$	$V_{DS} = -10V$, $I_D = -3A$	0.6	1.0	1.3	S
Input Capacitance	C_{iss}	$V_{GS} = -5V$, $f = 1MHz$	−	900	−	pF
Output Capacitance	C_{oss}	$V_{GS}=5V$, $V_{DS}=-5V$, $f=1MHz$	−	400	−	pF
Reverse Transfer Capacitance	C_{rss}	$V_{GS} = 5V$, $f = 1MHz$	−	40	−	pF
Turn-on Time	t_{on}		−	25	−	ns
Turn-off Time	t_{off}	$V_{DD} = -20V$, $I_D = -2A$	−	15	−	ns
Storage Time	t_{stg}		−	9	−	ns

Courtesy Hitachi America, Ltd.

2SJ48/49/50
P-Channel Power MOSFETs

AREA OF SAFE OPERATION

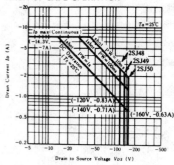

TYPICAL OUTPUT CHARACTERISTICS

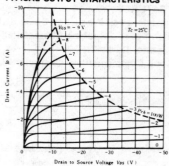

TYPICAL TRANSFER CHARACTERISTICS

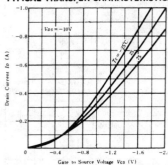

SWITCHING TIME VS. DRAIN CURRENT

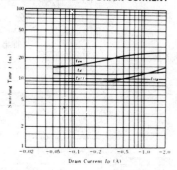

FORWARD TRANSFER ADMITTANCE VS. FREQUENCY

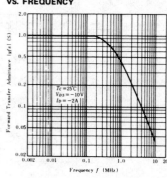

SWITCHING TIME TEST CIRCUIT

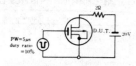

RESPONSE WAVE FORM

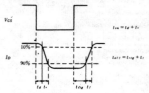

Courtesy Hitachi America, Ltd.

2SK133/134/135
N-Channel Power MOSFETs

■ FEATURES:
- ● **Superior High Frequency Characteristics**
- ● **High Speed Switching Characteristics**
- ● **Superior Durability**
- ● **Good Complementary Characteristics with 2SJ48/49/50**

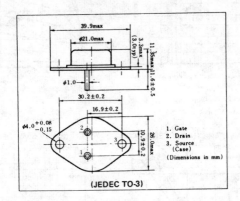

(JEDEC TO-3)

■ ABSOLUTE MAXIMUM RATINGS ($T_a = 25°C$)

Item	Symbol	Value	Unit
Drain to Source Voltage 2SK133 2SK134 2SK135	V_{DSX}	 120 140 160	V
Gate to Source Voltage	V_{GSS}	±14	V
Drain Current	I_D	7	A
Channel Dissipation	$P_{ch}*$	100	W
Storage Temperature	T_{stg}	−55~+150	°C

* Value at $T_C = 25°C$

MAXIMUM CHANNEL DISSIPATION CURVE

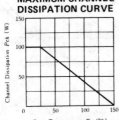

■ ELECTRICAL CHARACTERISTICS ($T_a = 25°C$)

Item	Symbol	Test Condition	min	typ	max	Unit
Drain to Source Breakdown Voltage 2SK133 2SK134 2SK135	$V_{(BR)DSX}$	$I_D = 10mA, V_{GS} = -10V$	 120 140 160	 − − −	 − − −	V
Gate to Source Breakdown Voltage	$V_{(BR)GSS}$	$I_G = ±100\mu A, V_{DS} = 0$	±14	−	−	V
Gate to Source Cut-off Voltage	$V_{GS(off)}$	$I_D = 100mA, V_{DS} = 10V$	0	1.0	1.5	V
Drain to Source Saturation Voltage	$V_{DS(sat)}$	$I_D = 7A, V_{GD} = 0$	−	−	12	V
Forward Transfer Admittance	$|y_{fs}|$	$V_{DS} = 10V, I_D \doteqdot 3A$	0.6	1.0	1.3	S
Input Capacitance	C_{iss}	$V_{GS} = 5V, f = 1MHz$	−	600	−	pF
Output Capacitance	C_{oss}	$V_{GS} = -5V, V_{DS}=5V, f=1MHz$	−	350	−	pF
Reverse Transfer Capacitance	C_{rss}	$V_{GS} = -5V, f = 1MHz$	−	10	−	pF
Turn-on Time	t_{on}		−	18	−	ns
Turn-off Time	t_{off}	$V_{DD} = 20V, I_D = 2A$	−	7	−	ns
Storage Time	t_{stg}		−	7	−	ns

Courtesy Hitachi America, Ltd.

2SK133/134/135
N-Channel Power MOSFETs

AREA OF SAFE OPERATION

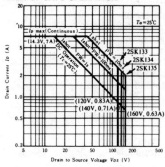

TYPICAL OUTPUT CHARACTERISTICS

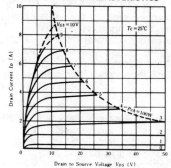

TYPICAL TRANSFER CHARACTERISTICS

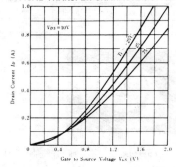

SWITCHING TIME VS. DRAIN CURRENT

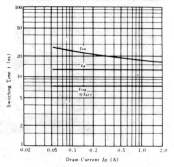

FORWARD TRANSFER ADMITTANCE
VS. FREQUENCY

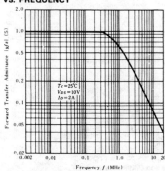

SWITCHING TIME TEST CIRCUIT

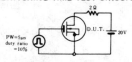

RESPONSE WAVE FORM

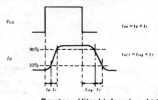

Courtesy Hitachi America, Ltd.

141

IRF100/101
High-Current N-Channel MOSFETs

HIGH CURRENT FIELD EFFECT TRANSISTORS

Advanced product design is made possible by utilizing the unique characteristics of Power MOSFET transistors. Control circuitry is simplified because MOSFETS are voltage controlled, second breakdown is eliminated, high current levels may be switched in nanoseconds, paralleling is easy, and temperature stability of the device parameters is outstanding.

The International Rectifier high current family of MOSFETS is particularly well suited for applications such as low voltage switching and linear power supplies, audio amplifiers, DC motor control, inverters, choppers, and IC and microprocessor-compatible power interface devices.

FEATURES:

- Fast Switching
- Low Drive Current
- No Second Breakdown
- Ease of Paralleling
- Excellent Temperature Stability

PRODUCT SUMMARY

	IRF100	IRF101
V_{DS}	80V	60V
$R_{D (on)}$	0.2Ω	0.2Ω
I_D	16A	16A

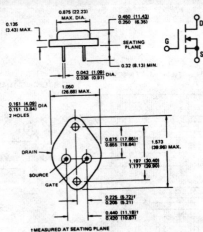

Conforms to JEDEC Outline TO-204AA (TO-3)
Dimensions in Inches and (Millimeters)

Courtesy International Rectifier Corp.

IRF100/101
High-Current N-Channel MOSFETs

ABSOLUTE MAXIMUM RATINGS

	Parameter	IRF100	IRF101	Units
V_{DS}	Drain-Source Voltage	80	60	V
V_{DGR}	Drain-Gate Voltage (R = 1 MΩ)	80	60	V
I_D	Continuous Drain Current	16	16	A
I_p	Pulsed Drain Current	32	32	A
V_{GS}	Gate-Source Voltage	±20		V
P_D	Max. Power Dissipation	125 (See Fig. 11)		W
	Linear Derating Factor	1.0 (See Fig. 11)		W/°C
$I_{L (peak)}$	Inductive Current, Clamped	32 (See Fig. 1 and 2) L = 100 μHy		A
T_J	Operating and			
T_{stg}	Storage Temperature Range	-55 to 150		°C
	Lead Temperature	300 (1.6mm from case for 10 sec)		°C

ELECTRICAL CHARACTERISTICS @ T_C = 25°C (UNLESS OTHERWISE SPECIFIED)

	Parameter	Type	Min.	Typ.	Max.	Units	Test Conditions
B_{VDSS}	Drain-Source Breakdown	IRF100	80	–	–	V	$V_{GS} = 0$
		IRF101	60	–	–	V	$I_D = 1.0$ mA
$V_{GS(th)}$	Gate Threshold Voltage	ALL	1	–	3	V	$V_{DS} = V_{GS}$, $I_D = 1$ mA
I_{GSS}	Gate-Body Leakage	ALL	–	–	100	nA	$V_{GS} = 20V$
I_{DSS}	Zero Gate Voltage Drain Current	ALL	–	0.1	1.0	mA	V_{DS} = Max. Rating, $V_{GS} = 0$
			–	0.2	4.0	mA	V_{DS} = 0.8 Max. Rating, $V_{GS} = 0$ $T_A = 125°C$.
$I_{D (on)}$	On-State Drain Current	ALL	16	–	–	A	$V_{DS} = 25V$, $V_{GS} = 10$
$R_{DS(ON)}$ and $r_{ds(on)}$	Static and Small Signal Drain-Source On State Resistance	ALL	–	0.16	0.2	Ω	$V_{GS} = 10V$, $I_D = 8A$
g_{fs}	Forward Transconductance	ALL	2.4	3.0	–	℧	$V_{DS} = 25$, $I_D = 8A$
C_{iss}	Input Capacitance	ALL	–	900	1200	pF	$V_{GS} = 0$, $V_{DS} = 25V$, f = 1.0 MHz
C_{oss}	Comm. Source Output	ALL	–	650	900	pF	See Fig. 10
C_{rss}	Reverse Transfer Capacitance	ALL	–	25	50	pF	
$t_{d(on)}$	Turn-On Delay Time	ALL	–	40	60	ns	$V_{DS} = 25V$, $I_D = 8A$
t_r	Rise Time	ALL	–	80	150	ns	(See Figs. 12 and 13)
$t_{d(off)}$	Turn-Off Delay Time	ALL	–	40	80	ns	$T_A = 125°C$ (MOSFET switching times are essentially independent
t_f	Fall Time	ALL	–	80	150	ns	of operating temperature.)

THERMAL CHARACTERISTICS

$R_{\theta JC}$	Maximum Thermal Resistance Junction-to-Case	ALL		1.0	°C/W

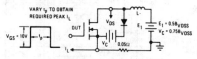

Fig. 1 — Clamped Inductive Test Circuit

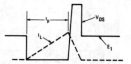

Fig. 2 — Clamped Inductive Waveforms

Courtesy International Rectifier, Corp.

IRF100/101
High-Current N-Channel MOSFETs

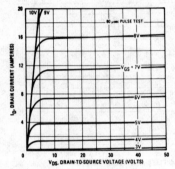

Fig. 3 — Output Characteristics

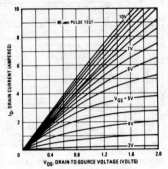

Fig. 4 — Saturation Characteristics

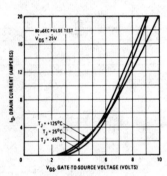

Fig. 5 — Transfer Characteristics

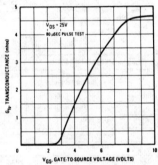

Fig. 6 — Transconductance Vs. Gate-to-Source Voltage

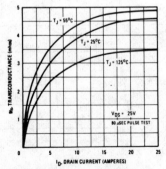

Fig. 7 — Transconductance Vs. Drain Current

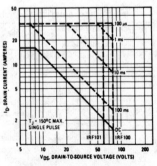

Fig. 8 — Maximum Safe Operating Area

Courtesy International Rectifier Corp.

IRF100/101
High-Current N-Channel MOSFETs

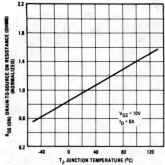

Fig. 9 — Normalized On-Resistance Vs. Temperature

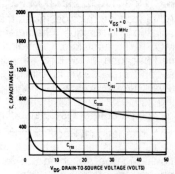

Fig. 10 — Capacitance Vs. Drain-to-Source Voltage

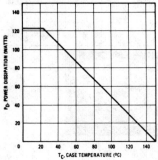

Fig. 11 — Power Vs. Temperature Derating

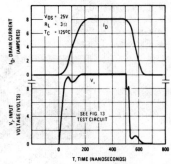

Fig. 12 — Switching Waveforms

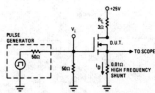

Fig. 13 — Switching Time Test Circuit

Courtesy International Rectifier Corp.

IRF300/301/305/306
High-Voltage N-Channel MOSFETs

HIGH VOLTAGE FIELD EFFECT TRANSISTORS

Advanced product design is made possible by utilizing the unique characteristics of Power MOSFET transistors. Control circuitry is simplified because MOSFETS are voltage controlled, second breakdown is eliminated, high current levels may be switched in nanoseconds, paralleling is easy, and temperature stability of the device parameters is outstanding.

The International Rectifier high voltage family of MOSFETS is particularly well suited for applications such as switching power supplies, motor control, inverters, converters, choppers, and high energy pulse circuits.

FEATURES:

- Fast Switching
- Low Drive Current
- No Second Breakdown
- Ease of Paralleling
- Excellent Temperature Stability

PRODUCT SUMMARY

	IRF300	IRF301	IRF305	IRF306
V_{DS}	400V	350V	400V	350V
$R_{D(on)}$	1.5Ω	1.5Ω	1.0Ω	1.0Ω
I_D	4A	4A	5A	5A

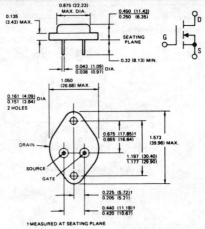

Conforms to JEDEC Outline TO-204AA (TO-3)
Dimensions in Inches and (Millimeters)

Courtesy International Rectifier Corp.

IRF300/301/305/306
High-Voltage N-Channel MOSFETs

ABSOLUTE MAXIMUM RATINGS

	Parameter	IRF300	IRF301	IRF305	IRF306	Units
V_{DS}	Drain — Source Voltage	400	350	400	350	V
V_{DGR}	Drain — Gate Voltage $^{(R = 1 M\Omega)}$	400	350	400	350	V
I_D	Continuous Drain Current	4		5		A
I_p	Pulsed Drain Current	8		10		A
V_{GS}	Gate — Source Voltage	±20				V
P_D	Max. Power Dissipation	125 (See Fig. 11)				W
	Linear Derating Factor	1.0 (See Fig. 11)				W/°C
$I_{L(peak)}$	Inductive Current, Clamped	10 (See Fig, 1 and 2) L = 100 μHy				A
T_J and T_{stg}	Operating and Storage Temperature Range	–55 to 150				°C
	Lead Temperature	300 (1.6mm from case for 10 sec)				°C

ELECTRICAL CHARACTERISTICS @ T_C = 25°C (UNLESS OTHERWISE SPECIFIED)

	Parameter	Type	Min.	Typ.	Max.	Units	Conditions
B_{VDSS}	Drain — Source Breakdown	IRF300 IRF305	400			V	$V_{GS} = 0$
		IRF301 IRF306	350			V	I_D = 1.0 mA
$V_{GS(th)}$	Gate Threshold Voltage	ALL	1		3	V	$V_{DS} = V_{GS}$, I_D = 1 mA
I_{GSS}	Gate — Body Leakage	ALL			100	nA	V_{GS} = 20V
I_{DSS}	Zero Gate Voltage Drain Current	ALL		0.1	1.0	mA	V_{DS} = Max. Rating, V_{GS} = 0
				0.2	4.0	mA	V_{DS} = 0.8 Max. Rating, V_{GS} = 0, T_A = 125°C
I_D (on)	On-State Drain Current	IRF300/1	4.0			A	V_{DS} = 25V, V_{GS} = 10V
		IRF305/6	5.0			A	
$R_{DS (ON)}$ and $r_{ds (on)}$	Static and Small-Signal Drain-Source On State Resistance	IRF300 IRF301		1.3	1.5	Ω	Static: V_{GS} = 10V, I_D = 2A Dynamic: V_{GS} = 10V, I_D = 0, f = 1 KHz
		IRF305 IRF306		0.8	1.0	Ω	Static: V_{GS} = 10V, I_D = 2.5 Dynamic: V_{GS} = 10V, I_D = 0, f = 1 KHz
g_{fs}	Forward Transconductance	ALL	1.0	2.5		$\mho$	V_{DS} = 25V, I_D = 2A
C_{iss}	Input Capacitance	ALL		800	1000	pF	V_{GS} = 0, V_{DS} = 25V, f = 1.0 MHz
C_{oss}	Comm. Source Output	ALL		350	500	pF	(See Fig. 10)
C_{rss}	Reverse Transfer Capacitance	ALL		15	20	pF	
$t_{d (on)}$	Turn-On Delay Time	ALL		30	50	nS	V_{DS} = 25V, I_D = 2A
t_r	Rise Time	ALL		50	100	nS	(See Figs. 12 and 13)
$t_{d (off)}$	Turn-Off Delay Time	ALL		80	150	nS	T_A = 125°C (MOSFET Switching times are essentially independent of operating temperature.)
t_f	Fall Time	ALL		50	100	nS	

THERMAL CHARACTERISTICS

$R_{\theta JC}$	Maximum Thermal Resistance Junction-to-Case	1.0	°C/W	

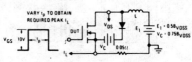

Fig. 1 — Clamped Inductive Test Circuit

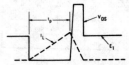

Fig. 2 — Clamped Inductive Waveforms

Courtesy International Rectifier Corp.

IRF300/301/305/306
High-Voltage N-Channel MOSFETs

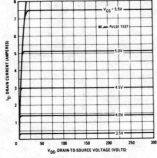

Fig. 3 — Output Characteristics

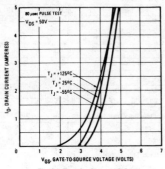

Fig. 4 — Transfer Characteristics

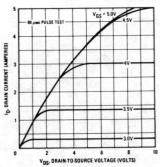

Fig. 5 — Saturation Characteristics (IRF300, IRF301)

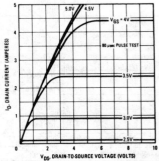

Fig. 6 — Saturation Characteristics (IRF305, IRF306)

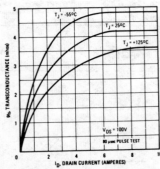

Fig. 7 — Transconductance Vs. Drain Current

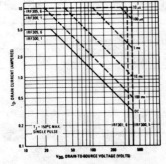

Fig. 8 — Maximum Safe Operating Area

Courtesy International Rectifier Corp.

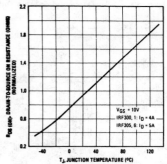

Fig. 9 — Normalized On-Resistance Vs. Temperature

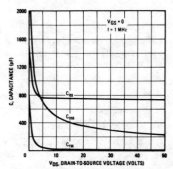

Fig. 10 — Capacitance Vs. Drain-to-Source Voltage

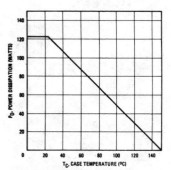

Fig. 11 — Power Vs. Temperature Derating

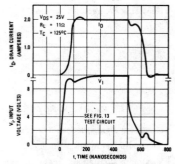

Fig. 12 — Switching Waveforms

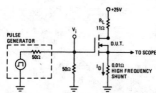

Fig. 13 — Switching Time Test Circuit

Courtesy International Rectifier Corp.

BD 512
P-Channel Power MOSFET

P-Channel Enhancement-Mode Power MOSFET

Normally off power MOSFET designed for
applications needing high input impedance
and fast switching times.

High input impedance

High speed switching

High power gain

No minority carrier storage time

CMOS Logic compatible input

No thermal runaway

No secondary breakdown

Zener protected gate to withstand overload

Paralleling is simple

Direct drive of inductive load

Heat sink connected to drain

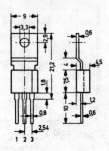

Fig. 1: Plastic case similar to
34 A 3 according to DIN 41 869

Weight approximately 1.5 g
Dimensions in mm

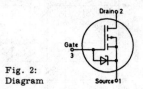

Fig. 2:
Diagram

ABSOLUTE MAXIMUM RATINGS

Maximum Drain-Source Voltage	$-V_{DSS}$	60	V
Maximum Drain-Gate Voltage	$-V_{DGS}$	60	V
Maximum Continuous Drain Current	$-I_D$	1.5	A
Maximum Gate (Zener) Current	$-I_G$	10	mA
Maximum Gate (Zener) Voltage	$-V_G$	15	V
Maximum Power Dissipation at 25 °C Case Temperature	P_{tot}	10	W
at 25 °C Free Air Temperature	P_{tot}	1.75	W
Thermal Resistance Junction to Heat Sink	R_{thS}	12.5	°C/W
Junction to Ambient Air	R_{thA}	70	°C/W
Temperature (Operating and Storage)	T_j, T_S	-55 to +150	°C

Courtesy ITT Semiconductors

BD 512
P-Channel Power MOSFET

ELECTRICAL CHARACTERISTICS at 25 °C

Symbol	Static Characteristics	Min.	Typ.	Max.	Unit	Test Conditions
$-BV_{DSS}$	Drain-Source Breakdown	60			V	$V_{GS} = 0$; $-I_D = 100\,\mu A$
$-V_{GS(th)}$	Gate Threshold Voltage	1.0		3.0 ·	V	$V_{GS} = V_{DS}$; $-I_D = 1\,mA$
$-I_{GSS}$	Gate-Body Leakage			0.5	μA	$-V_{GS} = 15$; $V_{DS} = 0$
$-I_{D(off)}$	Drain Cutoff Current			0.5	μA	$V_{GS} = 0$; $-V_{DS} = 25\,V$
$r_{DS(on)}$	Drain-Source ON Resistance*		6.5	10	Ω	$-V_{GS} = 10\,V$; $-I_D = 1\,A$
	Dynamic Characteristics					
g_m	Forward Transconductance*			150	m℧	$-V_{DS} = 10\,V$; $-I_D = 0.5\,A$
C_{iss}	Input Capacitance		50		pF	$V_{GS} = 0$; $-V_{DS} = 10\,V$; $f = 1\,MHz$
t_{ON}	Turn ON Time		4	10	ns	Test circuit see below
t_{OFF}	Turn OFF Time		4	10	ns	Test circuit see below

*Pulse Test Width - 80 µs; Pulse Duty Factor 1%

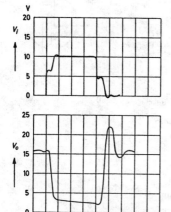

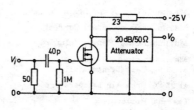

Fig. 3: Switching Performance of
Input and Output Voltages

Fig. 4: Test Circuit for the
Switching Times

BD 522
N-Channel Power MOSFET

N-Channel Enhancement-Mode Power MOSFET

Normally off power MOSFET designed for applications needing high input impedance and fast switching times.

High input impedance

High speed switching

High power gain

No minority carrier storage time

CMOS Logic compatible input

No thermal runaway

No secondary breakdown

Zener protected gate to withstand overload

Paralleling is simple

Direct drive of inductive load

Heat sink connected to drain

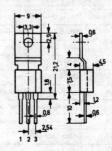

Fig. 1: Plastic case similar to 34 A 3 according to DIN 41 869

Weight approximately 1.5 g
Dimensions in mm

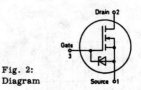

Fig. 2:
Diagram

ABSOLUTE MAXIMUM RATINGS

Maximum Drain-Source Voltage	V_{DSS}	60	V
Maximum Drain-Gate Voltage	V_{DGS}	60	V
Maximum Continuous Drain Current	I_D	1.5	A
Maximum Gate (Zener) Current	I_G	10	mA
Maximum Gate (Zener) Voltage	V_G	15	V
Maximum Power Dissipation at 25 °C Case Temperature	P_{tot}	10	W
at 25 °C Free Air Temperature	P_{tot}	1.75	W
Thermal Resistance Junction to Heat Sink	R_{thS}	12.5	°C/W
Junction to Ambient Air	R_{thA}	70	°C/W
Temperature (Operating and Storage)	T_j, T_S	-55 to +150	°C

Courtesy ITT Semiconductors

BD 522
N-Channel Power MOSFET

ELECTRICAL CHARACTERISTICS at 25 °C

Symbol	Static Characteristics	Min.	Typ.	Max.	Unit	Test Conditions
BV_{DSS}	Drain-Source Breakdown	60			V	$V_{GS} = 0; I_D = 100\ \mu A$
$V_{GS(th)}$	Gate Threshold Voltage	0.8		2.0	V	$V_{GS} = V_{DS}; I_D = 1\ mA$
I_{GSS}	Gate-Body Leakage			0.5	μA	$V_{GS} = 15\ V; V_{DS} = 0$
$I_{D(off)}$	Drain Cutoff Current			0.5	μA	$V_{GS} = 0; V_{DS} = 25\ V$
$r_{DS(on)}$	Drain-Source ON Resistance[a]		2.5	4.0	Ω	$V_{GS} = 10\ V; I_D = 1\ A$
	Dynamic Characteristics					
g_m	Forward Transconductance[a]		270		m℧	$V_{DS} = 10\ V; I_D = 0.5\ A$
C_{iss}	Input Capacitance		35		pF	$V_{GS} = 0; V_{DS} = 10\ V; f = 1\ MHz$
t_{ON}	Turn ON Time		4	10	ns	Test circuit see below
t_{OFF}	Turn OFF Time		4	10	ns	Test circuit see below

[a]Pulse Test Width - 80 μs; Pulse Duty Factor 1%

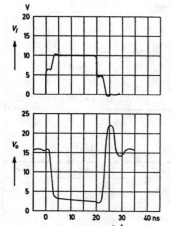

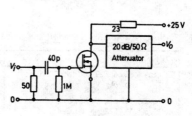

Fig. 3: Switching Performance of Input and Output Voltages

Fig. 4: Test Circuit for the Switching Times

Courtesy ITT Semiconductors

BS 170, BS 250
VMOSFETs

Development Sample Data

Enhancement Mode V-MOSFETs

BS 170: N-Channel Type

BS 250: P-Channel Type

Special Features:

High input impedance

High speed switching

No minority carrier storage time

CMOS logic compatible input

No thermal runaway

No secondary breakdown

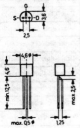

Plastic Package TO-92
Weight approx. 0.18 g
Dimensions in mm

Absolute Maximum Ratings

	Symbol	Value		Unit
		BS 170	BS 250	
Maximum Drain-Source Voltage	V_{DSS}	30	-30	V
Maximum Drain-Gate Voltage	V_{DGS}	30	-30	V
Maximum Continuous Drain Current	I_D	0.5	-0.5	A
Max. Power Dissipation at T_C = 25 °C	P_{tot}	0.6		W
Temperature (Operating and Storage)	T_j, T_S	-55 to + 150		°C

Static Characteristics BS 170 at T_j = 25 °C

	Symbol	Min.	Typ.	Max.	Unit
Drain-Source Breakdown Voltage at I_D = 100 μA, V_{GS} = 0	BV_{DSS}	30	-	-	V
Gate Threshold Voltage at V_{GS} = V_{DS}, I_D = 1 mA	$V_{GS(th)}$	0.5	-	2.5	V
Gate-Body Leakage Current at V_{GS} = 15 V, V_{DS} = 0	I_{GSS}	-	-	20	nA
Drain Cutoff Current at V_{DS} = 25 V, V_{GS} = 0	$I_{D(off)}$	-	-	0.5	μA
Drain-Source ON Resistance * at V_{GS} = 10 V, I_D = 0.2 A	$r_{DS(on)}$	-	2	3.5	Ω
Thermal Resistance Chip to Ambient Air	R_{thA}	-	125	-	°C/W

* Pulse Test Width - 80 μs; Pulse Duty Factor 1%.

Courtesy ITT Semiconductors

BS 170, BS 250
VMOSFETs

Dynamic Characteristics BS 170 at T_j = 25 °C

	Symbol	Min.	Typ.	Max.	Unit
Forward Transconductance * at V_{DS} = 10 V, I_D = 0.2 A, f = 1 MHz	g_m	-	110	-	mmho
Input Capacitance at V_{DS} = 10 V, V_{GS} = 0, f = 1 MHz	C_{iss}	-	20	-	pF
Turn ON Time at I_D = 0.2 A	t_{ON}	-	4	10	ns
Turn OFF Time at I_D = 0.2 A	t_{OFF}	-	4	10	ns

Static Characteristics of the BS 250 at T_j = 25 °C

	Symbol	Min.	Typ.	Max.	Unit
Drain-Source Breakdown Voltage at $-I_D$ = 100 µA, V_{GS} = 0	$-BV_{DSS}$	30	-	-	V
Gate Threshold Voltage at $V_{GS} = V_{DS}$, $-I_D$ = 1 mA	$-V_{GS(th)}$	0.5	-	2.5	V
Gate-Body Leakage Current at $-V_{GS}$ = 15 V, V_{DS} = 0	$-I_{GSS}$	-	-	20	nA
Drain Cutoff Current at $-V_{DS}$ = 25 V, V_{GS} = 0	$-I_{D(off)}$	-	-	0.5	µA
Drain-Source ON Resistance * at $-V_{GS}$ = 10 V, $-I_D$ = 0.2 A	$r_{DS(on)}$	-	4	7	Ω
Thermal Resistance Chip to Ambient Air	R_{thA}	-	125	-	°C/W

Dynamic Characteristics BS 250 at T_j = 25 °C

	Symbol	Min.	Typ.	Max.	Unit
Forward Transconductance * at $-V_{DS}$ = 10 V, $-I_D$ = 0.2 A, f = 1 MHz	g_m	-	80	-	mmho
Input Capacitance at $-V_{DS}$ = 10 V, V_{GS} = 0, f = 1 MHz	C_{iss}	-	20	-	pF
Turn ON Time at $-I_D$ = 0.2 A	t_{ON}	-	4	10	ns
Turn OFF Time at $-I_D$ = 0.2 A	t_{OFF}	-	4	10	ns

* Pulse Test Width - 80 µs; Pulse Duty Factor 1%.

Courtesy ITT Semiconductors

2N6656/57/58
N-Channel Power VMOSFETs

- **High Speed Switching**
- **CMOS to High Current Interface**
- **TTL to High Current Interface**
- **High Frequency Linear Amplifiers**
- **Line Drivers**
- **DC to DC Converters**
- **Switching Power Supplies**

BENEFITS

- Directly Interfaces to CMOS, TTL, DTL and MOS Logic Families
    Low Drive Current ($I_{GSS} < 10 \mu A$)
- Permits More Efficient and Compact Switching Designs
    Typical T_{ON} and $T_{OFF} < 5$ nsec
- Minimizes Component Count and Design Time/Effort
    Drives Inductive Loads Directly
    Fan Out From a CMOS Logic Gate > 100
    Easily Paralleled with Inherent Current Sharing Capability
    High Gain
- Improves Reliability
    Free From Secondary Breakdown Voltage Derating
    Inherently Reduces Output Current as Temperature Increases
    Input Protected From Static Discharge

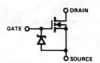

PIN	OUT
1	G
2	S
Case	Drain

ABSOLUTE MAXIMUM RATINGS

TO-3

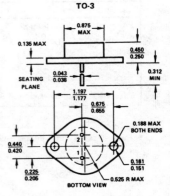

ALL DIMENSIONS IN INCHES

*Maximum Drain-Source Voltage
  2N6656. . . . . .35 V
  2N6657. . . . . .60 V
  2N6658. . . . . .90 V
*Maximum Drain-Gate Voltage
  2N6656. . . . . .35 V
  2N6657. . . . . .60 V
  2N6658. . . . . .90 V
*Maximum Continuous Drain Current . . . . 2.0 A
  Maximum Pulsed Drain Current . . . . 3.0 A
*Maximum Continuous Forward Gate Current. . . .2.0 mA[1]
*Maximum Pulsed Forward Gate Current. . . . . 100 mA
*Maximum Continuous Reverse Gate Current. . . . . . 100 mA
*Maximum Forward Gate-Source (Zener) Voltage . . . . .15 V
*Maximum Reverse Gate-Source Voltage . . . . 0.3 V
*Maximum Dissipation at 25°C Case Temperature . . . . .25 W
*Linear Derating Factor . . . . 200 mW/°C
*Temperature (Operating and Storage) . . . . . . −55 to +150°C
*Lead Temperature
  (1/16″ from case for 10 sec). . . . . .300°C

* Indicates JEDEC registered data

NOTE:
1. See graph, page 3.

Courtesy SEMTECH Corp.

2N6656/57/58
N-Channel Power VMOSFETs

ELECTRICAL CHARACTERISTICS (25°C unless otherwise noted)

#		Characteristic	2N6656 Min	2N6656 Typ	2N6656 Max	2N6657 Min	2N6657 Typ	2N6657 Max	2N6658 Min	2N6658 Typ	2N6658 Max	Unit	Test Conditions
1		BV$_{DSS}$ Drain-Source Breakdown	35			60			90			V	V$_{GS}$ = 0, I$_D$ = 10 µA
2			35			60			90				V$_{GS}$ = 0, I$_D$ = 2.5 mA
3*		V$_{GS(th)}$ Gate Threshold Voltage	0.8		2.0	0.8		2.0	0.8		2.0	V	V$_{DS}$ = V$_{GS}$, I$_D$ = 1 mA
4*		I$_{GSS}$ Gate Body Leakage		0.5	100		0.5	100		0.5	100	nA	V$_{GS}$ = 15 V, V$_{DS}$ = 0
5*					500			500			500		V$_{GS}$ = 15 V, V$_{DS}$ = 0, T$_A$ = 125 C (Note 2)
6*		I$_{DSS}$ Zero Gate Voltage Drain Current			10			10			10	nA	V$_{DS}$ = Max. Rating, V$_{GS}$ = 0
7*	S T A T I C				500			500			500	µA	V$_{DS}$ = 0.8 Max. Rating, V$_{GS}$ = 0, T$_A$ = 125 C (Note 2)
8				100			100			100		nA	V$_{DS}$ = 25 V, V$_{GS}$ = 0
9*		I$_{D(on)}$ ON-State Drain Current (Note 1)	1.0	2		1.0	2		1.0	2		A	V$_{DS}$ = 25 V, V$_{GS}$ = 10 V
10		V$_{DS(on)}$ Drain-Source Saturation Voltage (Note 1)		0.3			0.3			0.4		V	V$_{GS}$ = 5 V, I$_D$ = 0.1 Amp
11				1.0	1.5		1.0	1.5		1.1	1.6		V$_{GS}$ = 5 V, I$_D$ = 0.3 Amp
12				0.9			0.9			1.3			V$_{GS}$ = 10 V, I$_D$ = 0.5 Amp
13*				1.6	1.8		2.0	3.0		3.0	4.0		V$_{GS}$ = 10 V, I$_D$ = 1.0 Amp
14		r$_{DS(on)}$ Static Drain-Source ON-State Resistance		1.6	1.8		2.0	3.0		3.0	4.0	Ω	V$_{GS}$ = 10 V, I$_D$ = 1.0 Amp
15*		r$_{ds(on)}$ Small-Signal Drain-Source ON-State Resistance		1.6	1.8		2.0	3.0		3.0	4.0	Ω	V$_{GS}$ = 10 V, I$_D$ = 0, f = 1 kHz
16		g$_{fs}$ Forward Transconductance (Note 1)	170	250		170	250		170	250		m℧	V$_{DS}$ = 24 V, I$_D$ = 0.5 Amp
17*	D Y N A M I C	C$_{iss}$ Input Capacitance (Note 2)			50			50			50	pF	V$_{GS}$ = 0, V$_{DS}$ = 24 V, f = 1.0 MHz
18*		C$_{ds}$ Drain-Source Capacitance (Note 2)			40			40			40		V$_{GS}$ = 0, V$_{DS}$ = 24 V, f = 1.0 MHz
19		C$_{rss}$ Reverse Transfer Capacitance (Note 2)			10			10			10		V$_{GS}$ = 0, V$_{DS}$ = 24 V, f = 1.0 MHz
20*					35			35			35		V$_{GS}$ = 0, V$_{DS}$ = 0, f = 1.0 MHz
21*		t$_{d(on)}$ Turn-ON Delay Time (Note 2)		2	5		2	5		2	5	ns	See Switching Time Test Circuit
22*		t$_r$ Rise Time (Note 2)		2	5		2	5		2	5		
23*		t$_{d(off)}$ Turn-OFF Delay Time (Note 2)		2	5		2	5		2	5		
24*		t$_f$ Fall Time (Note 2)		2	5		2	5		2	5		

*Indicates JEDEC registered data

NOTES:
1. Pulse test—80 µsec pulse, 1% duty cycle.
2. Sample test.

TYPICAL PERFORMANCE CURVES (25°C unless otherwise noted)

Output Characteristics

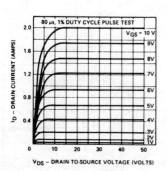

Saturation Characteristics

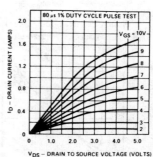

TYPICAL PERFORMANCE CURVES (Cont'd) (25°C unless otherwise noted)

Transfer Characteristic

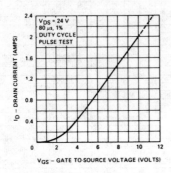

Drain-to-Source ON Resistance vs
Gate-to-Source Voltage

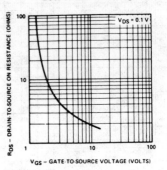

Transconductance vs Drain Current

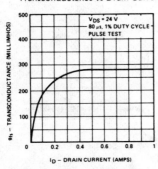

Transconductance vs
Gate-to-Source Voltage

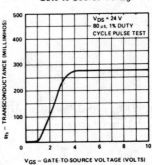

Normalized Drain-to-Source ON
Resistance vs Temperature

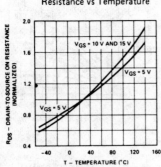

Capacitance vs
Drain-to-Source Voltage

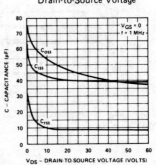

Courtesy SEMTECH Corp.

TYPICAL PERFORMANCE CURVES (Cont'd) (25°C unless otherwise noted)

Typical Zener
Breakdown Characteristic

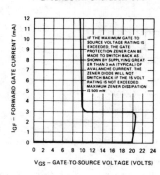

Output Conductance vs
Drain Current

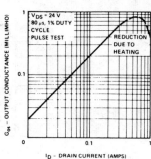

Power Dissipation vs
Case Temperature

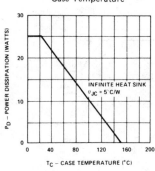

Maximum Safe Operating Region

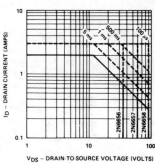

Thermal Response

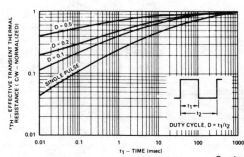

Courtesy SEMTECH Corp.

2N6656/57/58
N-Channel Power VMOSFETs

TYPICAL PERFORMANCE CURVES (Cont'd) (25°C unless otherwise noted)

Switching Waveforms

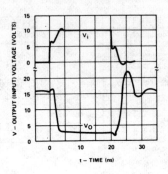

Switching Time Test Waveforms

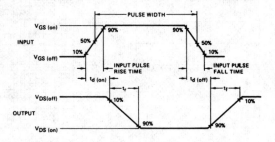

Switching Time Test Circuit

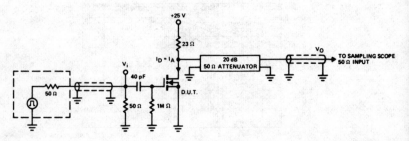

2N6656/57/58
N-Channel Power VMOSFETs

APPLICATIONS

CMOS Compatible Switch

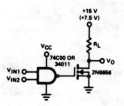

CMOS Switching Waveforms

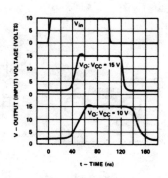
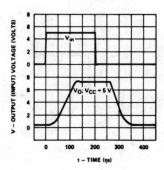

*High Speed CMOS Compatible Switch

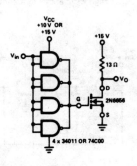

High Speed CMOS Switch Performance

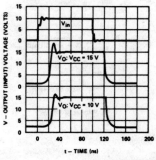

Courtesy SEMTECH Corp.

2N6656/57/58
N-Channel Power VMOSFETs

APPLICATIONS (Cont'd)

TTL Logic Compatible LED Driver

TTL Logic Compatible High Current
Solenoid Driver

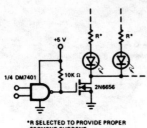

*R SELECTED TO PROVIDE PROPER
SEGMENT CURRENT

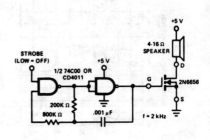

High Speed Open Collector
TTL Interface

Audio Alarm

Unidirectional Analog Switch

High Efficiency Light Dimmer

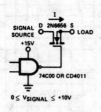

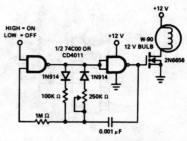

Courtesy SEMTECH Corp.

162

2N6656/57/58
N-Channel Power VMOSFETs

APPLICATIONS (Cont'd)

High Power Bidirectional Analog Switch

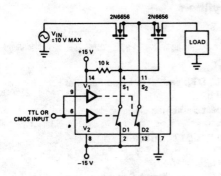

Parallel Operation

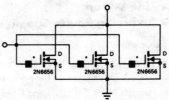

*TO PREVENT SPURIOUS OSCILLATIONS, A 500 Ω-
1K Ω RESISTOR OR FERRITE BEAD (FOR HIGHER
SPEED) SHOULD BE CONNECTED IN SERIES WITH
EACH GATE.

Series Operation

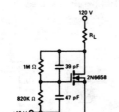

Microprocessor to
High Power Interface

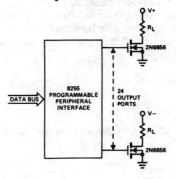

Low Level to High
Current Interface

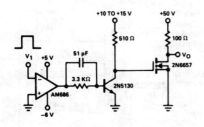

Performance of the
High Current Interface

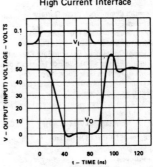

Courtesy SEMTECH Corp.

163

2N6659/60/61
N-Channel Power VMOSFETs

- **High Speed Switching**
- **CMOS to High Current Interface**
- **TTL to High Current Interface**
- **High Frequency Linear Amplifiers**
- **Line Drivers**
- **Switching Power Supplies**

BENEFITS

- Directly Interfaces to CMOS, TTL, DTL and MOS Logic Families
  Low Drive Current ($I_{GSS} < 10\ \mu A$)
- Permits More Efficient and Compact Switching Designs
  Typical T_{ON} and $T_{OFF} < 5$ nsec
- Minimizes Component Count and Design Time/Effort
  Drives Inductive Loads Directly
  Fan Out From a CMOS Logic Gate > 100
  Easily Paralleled with Inherent Current Sharing Capability
  High Gain
- Improves Reliability
  Free From Secondary Breakdown Voltage Derating
  Inherently Reduces Output Current as Temperature Increases
  Input Protected from Static Discharge

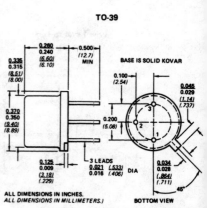

PIN	OUT
1	S
2	G
3	Drain & Case

TO-39

ABSOLUTE MAXIMUM RATINGS

* Maximum Drain-Source Voltage
  2N6659 . . . . 35 V
  2N6660 . . . . 60 V
  2N6661 . . . . 90 V
* Maximum Drain-Gate Voltage
  2N6659 . . . . 35 V
  2N6660 . . . . 60 V
  2N6661 . . . . 90 V
* Maximum Continuous Drain Current . . . . 2.0 A
  Maximum Pulsed Drain Current . . . . 3.0 A
* Maximum Continuous Forward Gate Current . 2.0 mA(1)
* Maximum Pulsed Forward Gate Current . . . . . . 100 mA
* Maximum Continuous Reverse Gate Current . . . 100 mA
* Maximum Forward Gate-Source (Zener) Voltage . . 15 V
* Maximum Reverse Gate-Source Voltage . . . . 0.3 V
* Maximum Dissipation at 25°C Case Temperature 6.25 W
* Linear Derating Factor . . . . 50 mW/°C
* Temperature (Operating and Storage) . . . −55 to+150°C
* Lead Temperature
  (1/16″ from case for 10 seconds) . . . . 300°C

* Indicates JEDEC registered data

NOTE:
1. See graph, page 3.

ALL DIMENSIONS IN INCHES.
ALL DIMENSIONS IN MILLIMETERS.)

BOTTOM VIEW

Courtesy SEMTECH Corp.

164

2N6659/60/61
N-Channel Power VMOSFETs

*ELECTRICAL CHARACTERISTICS (25°C unless otherwise noted)

	Symbol	Characteristic	2N6659 Min	Typ	Max	2N6660 Min	Typ	Max	2N6661 Min	Typ	Max	Unit	Test Conditions
1	BV$_{DSS}$	Drain-Source Breakdown	35			60			90			V	V$_{GS}$ = 0, I$_D$ = 10 µA
2			35			60			90			V	V$_{GS}$ = 0, I$_D$ = 2.5 mA
3*	V$_{GS(th)}$	Gate Threshold Voltage	0.8			0.8		2.0	0.8		2.0	V	V$_{DS}$ = V$_{GS}$, I$_D$ = 1 mA
4*	I$_{GSS}$	Gate-Body Leakage		0.5	100		0.5	100		0.5	100	nA	V$_{GS}$ = 15 V, V$_{DS}$ = 0
5*					500			500			500		V$_{GS}$ = 15 V, V$_{DS}$ = 0, T$_A$ = 125°C (Note 2)
6*	I$_{DSS}$	Zero Gate Voltage Drain Current			10			10			10	µA	V$_{DS}$ = Max. Rating, V$_{GS}$ = 0
7*					500			500			500		V$_{DS}$ = 0.80 Max. Rating, V$_{GS}$ = 0, T$_A$ = 125°C (Note 2)
8					100			100			100	nA	V$_{DS}$ = 25 V, V$_{GS}$ = 0
9*	I$_{D(on)}$	ON-State Drain Current (Note 1)	1.0	2		1.0	2		1.0	2		A	V$_{DS}$ = 25 V, V$_{GS}$ = 10 V
10	V$_{DS(on)}$	Drain-Source Saturation Voltage (Note 1)		0.3			0.3			0.4		V	V$_{GS}$ = 5 V, I$_D$ = 0.1 Amp
11				1.0	1.5		1.0	1.5		1.1	1.6		V$_{GS}$ = 5 V, I$_D$ = 0.3 Amp
12				0.9			0.9			1.3			V$_{GS}$ = 10 V, I$_D$ = 0.5 Amp
13*				1.6	1.8		2.0	3.0		3.0	4.0		V$_{GS}$ = 10 V, I$_D$ = 1.0 Amp
14*	r$_{DS(on)}$	Static Drain-Source ON-State Resistance (Note 1)		1.6	1.8		2.0	3.0		3.0	4.0	Ω	V$_{GS}$ = 10 V, I$_D$ = 1.0 Amp
15*	r$_{ds(on)}$	Small-Signal Drain-Source ON-State Resistance		1.6	1.8		2.0	3.0		3.0	4.0		V$_{GS}$ = 10 V, I$_D$ = 0, f = 1 kHz
16	g$_{fs}$	Forward Transconductance (Note 1)	170	250		170	250		170	250		m℧	V$_{DS}$ = 24 V, I$_D$ = 0.5 Amp
17*	C$_{iss}$	Input Capacitance (Note 2)		50			50			50		pF	V$_{GS}$ = 0, V$_{DS}$ = 24 V, f = 1.0 MHz
18*	C$_{ds}$	Drain-Source Capacitance (Note 2)		40			40			40			
19*	C$_{rss}$	Reverse Transfer Capacitance (Note 2)		10			10			10			V$_{GS}$ = 0, V$_{DS}$ = 24 V, f = 1.0 MHz
20*				35			35			35			V$_{GS}$ = 0, V$_{DS}$ = 0, f = 1.0 MHz
21*	t$_{d(on)}$	Turn-ON Delay Time (Note 2)		2	5		2	5		2	5		
22*	t$_r$	Rise Time (Note 2)		2	5		2	5		2	5		
23*	t$_{d(off)}$	Turn-OFF Delay Time (Note 2)		2	5		2	5		2	5	ns	See Switching Time Test Circuit
24*	t$_f$	Fall Time (Note 2)		2	5		2	5		2	5		

*Indicates JEDEC registered data

NOTES:
1. Pulse test—80 µsec pulse, 1% duty cycle.
2. Sample test.

TYPICAL PERFORMANCE CURVES (25°C unless otherwise noted)

Output Characteristics

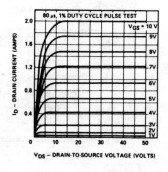

Saturation Characteristics

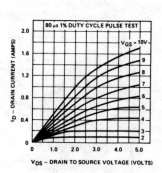

Courtesy SEMTECH Corp.

TYPICAL PERFORMANCE CURVES (Cont'd) (25°C unless otherwise noted)

Transfer Characteristic

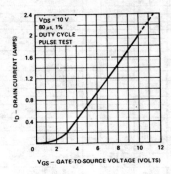

Drain-to-Source ON Resistance vs Gate-to-Source Voltage

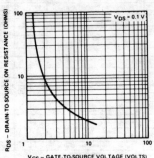

Transconductance vs Drain Current

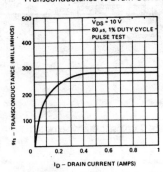

Transconductance vs Gate-to-Source Voltage

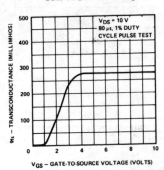

Normalized Drain-to-Source ON Resistance vs Temperature

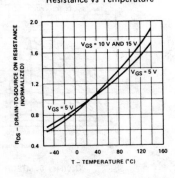

Capacitance vs Drain-to-Source Voltage

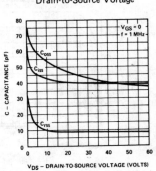

Courtesy SEMTECH Corp.

2N6659/60/61
N-Channel Power VMOSFETs

Typical Zener Breakdown Characteristic

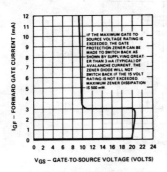

IF THE MAXIMUM GATE TO SOURCE VOLTAGE RATING IS EXCEEDED, THE GATE PROTECTION ZENER CAN BE MADE TO SWITCH BACK AS SHOWN BY SUPPLYING GREATER THAN 3 mA (TYPICAL) OF AVALANCHE CURRENT. THE ZENER DIODE WILL NOT SWITCH BACK IF THE 15 VOLT RATING IS NOT EXCEEDED. MAXIMUM ZENER DISSIPATION IS 500 mW.

Output Conductance vs Drain Current

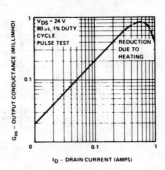

Power Dissipation vs Case or Ambient Temperature

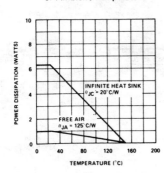

Maximum Safe Operating Region

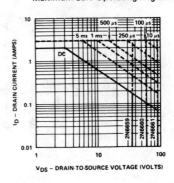

Switching Time Test Waveforms

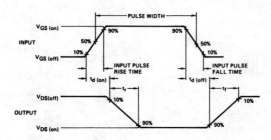

Courtesy SEMTECH Corp.

TYPICAL PERFORMANCE CURVES (Cont'd) (25°C unless otherwise noted)

Switching Waveforms

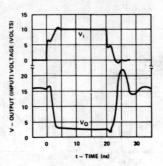

Switching Time Test Circuit

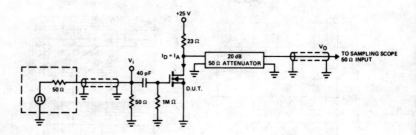

Courtesy SEMTECH Corp.

2N6659/60/61
N-Channel Power VMOSFETs

APPLICATIONS

CMOS Compatible Switch

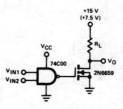

CMOS Switching Waveforms

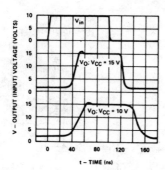 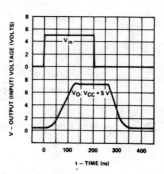

High Speed CMOS Compatible Switch

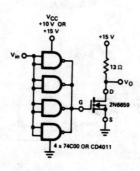

High Speed CMOS Switch Performance

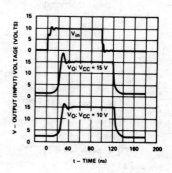

Courtesy SEMTECH Corp.

2N6659/60/61
N-Channel Power VMOSFETs

APPLICATIONS (Cont'd)

Parallel Operation

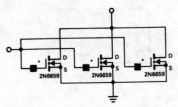

*TO PREVENT SPURIOUS OSCILLATIONS, A 500 Ω-
1K Ω RESISTOR OR FERRITE BEAD (FOR HIGHER
SPEED) SHOULD BE CONNECTED IN SERIES WITH
EACH GATE.

TTL Logic Compatible LED Driver

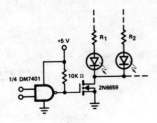

TTL Compatible High Speed
High Current Switch

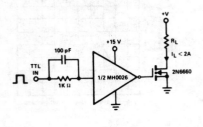

High Speed Open Collector
TTL Interface

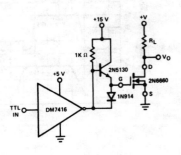

Series Operation

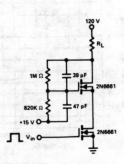

Audio Alarm

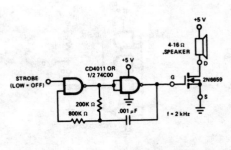

Courtesy SEMTECH Corp.

APPLICATIONS (Cont'd)

High Efficiency Light Dimmer

Unidirectional Analog Switch

Bidirectional Analog Switch

Laser Diode Pulser

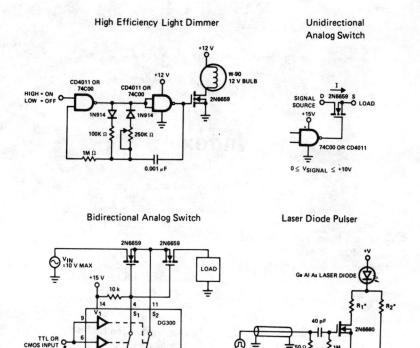

*R₁, R₂ SELECTED TO PROVIDE PROPER BIAS AND PULSE CURRENTS

High Current Interface

Performance of the High Current Interface

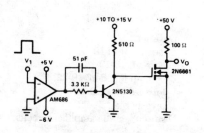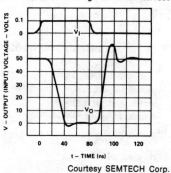

Courtesy SEMTECH Corp.

171

Index

A

Absolute maximum ratings, typical
    VMOS, 33-34
Ac load control, 86
Analog switching, 92-96
Applications
    linear, 11-14
    switching, 14-15
Audio amplifier circuits, 51-64
    Class D switching, 59-61
    Class G switching, 62-64
    distortion in, 51-52
    4-watt, 2 transistor, 52-54
    15-watt high-fidelity, 54-56
    40-watt high-fidelity, 56-57, 59
    100-watt ultralow distortion, 58, 59
Automotive ignition systems, 96-97

B

Basic circuit configurations, 38-50
    common-gate, 42-43
    common-source, 38-42
    parallel operation, 47
    series operation, 48-50
    source-follower, 43-44
Bias
    forward, 20-21
    reverse, 21-22

Bipolar junction transistor, 19
    operation of, 22-23

C

Capacitance, diffusion, 19
Charge carriers, 18-19
    majority, 22
    minority, 22
Common-gate circuit, 42-43
Common-source circuit, 38-42
Conductors, 17
Constant current source, 88-89
Core driver, magnetic, 85-86
Current
    definition of, 18-19
    diffusion, 18-19
    drift, 18
    hogging, 22

D

Diffusion
    capacitance, 19
    current, 18-19
Diode, np junction, 19-22
    forward bias, 20-21
    reverse bias, 21-22
Distortion, audio amplifier circuits,
    51-52

Drift current, 18
Dynamic electrical characteristics,
    typical VMOS, 34-36

E

ECL interface, 83
Electrical characteristics
    dynamic, typical VMOS, 34-36
    static, typical VMOS, 34

F

Forward bias, 20-21

H

HEXFET, 29
High current variable resistor, 90
Hogging, current, 22

I

Ignition systems, automotive, 96-97
Inductive loads, switching, 46-47
Input impedance, 25-26
Insulators, 17
Interfacing
    logic-level, 80-83
        ECL interface, 83
        TTL interfaces, 80-82
    peripheral, 84-87
        ac load control, 86
        lamp driver, 85
        LED driver, 86
        magnetic core driver, 85-86
        paper tape reader and stepping
            motor drive, 86-87
        relay or solenoid driver, 85
Inverter circuits, VMOS, 76-77

L

Lamp driver, 85
Large-scale integration devices, 98-102
    random access memories, 98-99
    read-only memory, 99
LED driver, 86

Linear
    applications, 11-14
    element, ideal, 10-11
    regulators, 77-78
Line driver circuit, 84
Logic-level interfacing, 80-83
    ECL interface, 83
    TTL interfaces, 80-82

M

Majority carriers, 22
Memories
    random access, 98-99
    read-only, 99
Microcomputer applications, 79-87
    line driver circuit, 84
    logic-level interfacing, 80-83
        ECL interface, 83
        TTL interfaces, 80-82
    peripheral interfacing, 84-87
        ac load control, 86
        lamp driver, 85
        LED driver, 86
        magnetic core driver, 85-86
        paper tape reader and stepping
            motor drive, 86-87
        relay or solenoid driver, 85
Minority carriers, 22

N

Np diode junction, 19-22
    forward bias, 20-21
    reverse bias, 21-22

O

Overload protection, 78

P

Parallel operation, VMOS devices, 47
Parameters, typical VMOS, 33-36
Peripheral interfacing, 84-87
    ac load control, 86
    lamp driver, 85
    LED driver, 86
    magnetic core driver, 85-86

Peripheral interfacing—cont
  paper tape reader and stepping
    motor drive, 86-87
  relay or solenoid driver, 85
Power control, 90-91
Power supply applications, 72-78
  inverter circuits, 76-77
  linear regulators, 77-78
  overload protection, 78
  switching regulators, 72-76
Pulse switching circuit, 91-92

R

Radio frequency applications, 65-71
  1.8- to 54-MHz, 16-watt broadband
    amplifier, 67-71
  40- to 220-MHz broadband
    amplifier, 66-67
Random access memories, 98-99
Read-only memory, 99
Regulators
  linear, 77-78
  switching, 72-76
Relay or solenoid driver, 85
Resistor, variable, high current, 90
Reverse bias, 21-22

S

Semiconductor operation, theory of,
  16-30
Series operation, VMOS devices, 48-50
Source-follower circuit, 43-44

Static electrical characteristics, typical
    VMOS, 34
Switch, ideal, 11
Switching
  analog, 92-96
  applications, 14-15
  considerations, 44-46
  inductive loads, 46-47
  regulators, 72-76
  speed, 75

T

Thermal runaway, 22
TTL interfaces, 80-82

V

Variable resistor, high current, 90
Vertical field-effect transistor, 23-25
VMOS
  advantages, 27-29
    circuit simplicity, 27-28
    reduced size requirements, 28-29
    self-protection, 29
    handling precautions, 31-32
    parallel operation, 47
    power considerations, 33
    protected versus unprotected
      gates, 36-37
    series operation, 48-50
    typical parameters, 33-36
  disadvantages, 26-27
  inverter circuits, 76-77

TO THE READER

This book is one of an expanding series of books that will cover the field of basic electronics and digital electronics from basic gates and flip-flops through microcomputers and digital telecommunications. We are attempting to develop a mailing list of individuals who would like to receive information on the series. We would be delighted to add your name to it if you would fill in the information below and mail this sheet to us. Thanks.

1. I have the following books:

_____

_____

_____

_____

_____

_____

_____

_____

_____

_____

_____

2. My occupation is: ☐ student   ☐ teacher, instructor   ☐ hobbyist

   ☐ housewife   ☐ scientist, engineer, doctor, etc.   ☐ businessman

   ☐ Other: _____

Name (print): _____

Address _____

City _____   State _____

Zip Code _____

Mail to:

   Books
   P.O. Box 715
   Blacksburg, Virginia 24060